DIGITAL ODYSSEY

TECH WHIZ GIRL BURSTS OWN BUBBLE

a memoir

SE QUINN

Ballast
Books

Ballast Books, LLC
www.ballastbooks.com

ISBN:
Hardcover: 978-1-966786-51-1
Paperback: 978-1-966786-49-8
eBook: 978-1-966786-52-8

Printed in the United States of America

Published by Ballast Books
www.ballastbooks.com

For more information, bulk orders, appearances, or speaking requests,
please email: info@ballastbooks.com

For Susie

CONTENTS

Ithaka

C. P. CAVAFY

As you set out for Ithaka
hope your road is a long one,
full of adventure, full of discovery.
Laistrygonians, Cyclops,
angry Poseidon—don't be afraid of them:
you'll never find things like that on your way
as long as you keep your thoughts raised high,
as long as a rare excitement
stirs your spirit and your body.
Laistrygonians, Cyclops,
wild Poseidon—you won't encounter them
unless you bring them along inside your soul,
unless your soul sets them up in front of you.

Hope your road is a long one.
May there be many summer mornings when,
with what pleasure, what joy,
you enter harbors you're seeing for the first time;
may you stop at Phoenician trading stations
to buy fine things,
mother of pearl and coral, amber and ebony,
sensual perfume of every kind—
as many sensual perfumes as you can;
and may you visit many Egyptian cities
to learn and go on learning from their scholars.

Keep Ithaka always in your mind.
Arriving there is what you're destined for.
But don't hurry the journey at all.
Better if it lasts for years,
so you're old by the time you reach the island,
wealthy with all you've gained on the way,
not expecting Ithaka to make you rich.

Ithaka gave you the marvelous journey.
Without her you wouldn't have set out.
She has nothing left to give you now.

And if you find her poor, Ithaka won't have fooled you.
Wise as you will have become, so full of experience,
you'll have understood by then what these Ithakas mean.

PROLOGUE

"I don't know much about being a millionaire,
but I'll bet I'd be darling at it."

Dorothy Parker

1996—PALO ALTO, CALIFORNIA

The invitation drops casually, like everyone's free after dark. But I was doing the math in my head—miles, timing, how far I could stretch a weeknight without knocking things off balance at home. Jake can order pizza, I reason. I've cooked every other night this week.

"We meet on Thursdays once a month—engineers, strategists, and start-up types, people thinking a few steps ahead. You should come," Jonas says, smiling.

It's a school night for Emma, but I think I can pull it off.

After all, Jonas is Jonas Hart—recently an editor of the *Harvard Law Review* and now working on digital tech policy at the FCC. He's helping shape the early frameworks for internet commerce. A few years from now, he'll launch a cross-industry forum exploring the future of connectivity and networks. He's on the board of advisors for my tech company, but

I haven't socialized with him. Running in his circle could open doors to rooms even grander than the ones I'm already constructing.

I think Jonas knows I have kids, but I present myself like one of the gang, as though it's nothing for me to gallivant all over Silicon Valley, San Francisco, and Marin County while running both a pioneering tech start-up and a popular women's networking organization, all with two little girls, aged five and one, at home.

On Thursday, instead of taking the high-speed catamaran commuter ferry from Tiburon into the city as usual, I drive my brand-new Jeep Cherokee over the Golden Gate Bridge to my office in a quaint brick building on the corner of Pacific and Sansome. It's a full day, but I know that if I leave the office by 4:00 p.m., I can make it through the city before I'm caught in the tedious rush hour crawl south. When I'm out the door on schedule, I feel confident. I have plenty of time. Well, maybe too much time. Dinner is scheduled for 9:00 p.m.

Once I pass the Cesar Chavez ramp, I-280 opens up. A sense of calm washes over me as the traffic thins and the highway expands, iconic green hills rolling out ahead of me, undulating like ocean swells catching the light of golden hour. It's where I grew up. The radio is on, and nineties alt-rock slips through my Jeep's speakers, providing a happy soundtrack as I speed across the scene.

"Hang tight, freaks and geeks. This is Big Rick Stuart, and that was Third Eye Blind with 'Semi-Charmed Life'—a little summertime sugar rush with a side of chaos from right here in San Francisco. You're listening to Live 105—modern rock for people who ditched the dress code and still made partner. Beck's up next, your Snapple's sweating, and it's officially too nice out to pretend you're working late."

My job demands constant focus, so I use moments like this to reset —to take a breath, scan the horizon, stop thinking about where I'm going, and register where I am. I purposely take in the view, look at the skyline, the way a foothill catches light. The nature of it fills me with something, *awareness*. The opportunity to bathe in the essence of a moment that asks for nothing of me. The deliberate act of looking around is a way of capturing a fleeting moment as it happens. I've noticed in the past that when

I reach a target, it isn't as shiny, as golden and brilliant as it was when it was still a dream. When I've accomplished a dream, I'm already bored of it somehow and striving for the next goal and the next after that. Taking moments to stop on the way to the goals I'd set was my secret gift to myself. A choice to *be*, not just *do*.

Every part of this life—I built. No magic. Just work. It's holding because I'm the one holding it up. There was no lucky break. This is forward motion I can trust—because it's mine, and I made it from scratch.

I'm very early to Palo Alto, so I duck into a café for some quiet time with my laptop. Inside, I engage the Ricochet modem slapped to the lid with Velcro and point its fat gray antenna upward. Then, I check my email and reply to various correspondences while sipping a perfect cappuccino with steamed whole milk and extra heavy cream, just the way I like it. I float in my chair in luxury and anticipation, uber cool, utterly content.

The sun has set, giving way to a balmy evening as I wait for the late dinner hour. Occasionally, I pop out to feed the parking meter, reluctantly abandoning my laptop and the back-to-back cappuccinos accumulating on the table, and when 8:30 finally arrives, I pack up my portable office and walk the block south to University Avenue.

I grow aware of my own presence as I nervously enter the posh and crowded hotel lobby where the restaurant is located. Tall, handsome people with bright smiles and confidence surround me. Unlike the San Francisco tech scene, with its grunge bike messenger rhythm, the Palo Alto crowd is taller, their suits more crisply pressed, their teeth seem shinier. Their vibe is not a soulful one but rather a more clinical, scholarly energy.

My black pencil skirt goes perfectly with my Manolo Blahnik slingback kitten heels, but I am awkward in the artfully cut white cotton origami blouse I have folded into the top of my skirt. *Your ensemble is smart and fashionable*, I remind myself, taking comfort in the thought. My collarbones show, demure but sexy, between the exaggerated lapel, but I worry that my light layer of liquid foundation will seep onto the oversized triangles of thick white fabric if I turn my head naturally to either side. The effort it takes me to appear effortless is significant.

I am chronically early as a rule, even when I try to be late, and once again I'm half an hour early for the reservation in a place where busy and important people are *never* early. I don't know anyone attending this dinner other than Jonas, so I take a stool at the corner of the leather bar and order a glass of rosé while I wait for him to arrive.

I get more uneasy as I overanalyze the social expectations of such a meetup while sitting here alone. What will I be expected to talk about? I'm always the only one my age with kids. What will the cool ones want to know about me to determine if I fit in? Will they ask who I know in the industry? Will there be a conversation about Frisbee tournaments? Certainly no one will discuss fashion—but maybe restaurants? Restaurants in Manhattan? Oh, I hope they don't talk a lot about university life. I never went, a dirty little secret that takes too long to explain.

My mind fills with more questions than solutions. Why am I so self-conscious approaching this dinner group when, in all the ways that matter, I have accomplished as much as any of them? They who matriculated to Ivy League grad schools, who are polymaths possessing or in the process of obtaining MBAs *and* law degrees from MIT and Harvard. Some of them are members of think tanks led by David Liddle, one of the visionaries at Xerox Parc, or shape policy at the Electronic Frontier Foundation with Esther Dyson. Some are junior partners at the top venture capital firms on Sand Hill Road. Do they ride horses? I feel like talking about my kids or my tomato garden, but I won't.

I recognize Jonas Hart as he finally strolls in, accompanied by three bright, fun, smart-looking women of various ethnicities, all around twenty-five, a couple of years younger than I am. Jonas, when he spots me, looks exuberantly pleased to see me here, and that puts me at ease.

The cohort leads me to a large table in the middle of this trendy, wood-and-leather-decorated hot spot. Boisterous laughter sparkles throughout the restaurant, and the dancing lights of University Avenue flicker outside the impressive two-story windows. As we approach the dinner table, I nod and raise one hand in a meager wave. Everyone turns their brightness and attention to me with huge, authentic smiles.

I feel welcome. It is a great time to be here. Right here. Right now. It's no wonder everyone beams. They're brimming with realized potential.

After we are all seated and have been served an opening glass of 1995 Duckhorn sauvignon blanc, Jonas clinks his glass with a spoon, quieting us. Our table turns to watch him, and other tables in the room, intrigued, turn in our direction as well.

"Everyone, if I may have your attention for a moment," he says over the lively clamor. "I have the distinct pleasure of introducing someone who is at the forefront of advancing real-time information streaming technology. Now, I could bore you with a list of her impressive accomplishments, but that's what search engines are for. Instead, let me just say that when I was asked to host this month's gathering, I wanted to invite someone who could light up this room, and I immediately thought of tonight's guest."

"Woot woot!" exclaims a strapping young man at the far end of the table.

"In a world of business suits and PowerPoint presentations, Susan manages to make innovation feel fresh and instinctive. You see, it's not every day you get to meet a visionary who can turn a brainstorm into a lightning storm. So, please give a warm welcome to the CEO and founder of Wordcasters.com, Susan Quinn."

The table comes alive with clapping and cheering. As the sound crashes over me, I know it's my turn to speak.

"Thank you so much for that incredibly warm welcome," I begin, standing and locking my hands together to show my gratitude. "It's truly an honor to be here among such an impressive group of individuals. And if turning brainstorms into lightning storms means we're in for an electrifying evening, I couldn't be more excited. Let's make tonight memorable, and I look forward to some great conversations. Thank you, Jonas, for the humbling intro and to all of you for having me."

I make funny faces as I talk for some reason. I always have. I can string together a series of words with adequate success on certain days, but my face still fusses about like a goofy cartoon character. I used to think it a betrayal, but I'm okay with it now. It's me.

As the murmur of voices resumes, a woman seated across the table leans in my direction. "Susan," she half shouts, "what's the next major development you're working on?"

I contemplate for a moment before replying.

"Keeping in mind bandwidth," I offer, "DSL and cable modems—they're coming, but not for everyone and not all at once. For now, it's all about products that combine communication technologies that are deliverable over the existing bandwidth and creating that interface so that as novel communications technologies become available for the mainstream, we've already got the user base plugged in to integrate seamlessly. That's what we're working on right now at Wordcasters."

At that, a spectacled junior associate at Draper, Fisher, Jurvetson leans in with a follow-up. "Do you have an elevator? I'm quite interested in this stage. Have you done a Series A?"

"Not yet. We're going for our Series A this summer."

"Nice! Could you share your pitch for Wordcasters and perhaps delve into a more detailed explanation of your company's core operations? I'm keen to understand the specifics of your work."

This, I know, is how I get invitations and insider referrals to meet with partners at the big venture capital firms. It's remarkably easy to land the meetings and make the pitch. What isn't easy for me is the fear and the doubt, but my desire for new experiences, while intimidating, always wins out.

From here, the conversation flows as smoothly as the wine—we transition to bolder reds and traverse topics from internet infrastructure and cybersecurity to e-commerce and privacy. I engage with passion, my words carried by the shared excitement of this exclusive gathering.

Soon, Jonas takes the proverbial microphone, and we all sit rapt as he sheds light on his work in internet privacy and online civil liberties. Jonas describes spearheading progress on a wide range of digital rights issues and how the work he is doing will play a significant role in advocating for and protecting individuals' rights in the digital age. As technology continues to develop at breakneck speed, safeguarding social justice, privacy, and equity will become a critical responsibility for us tech visionaries.

As plates are cleared and desserts arrive, the conversation shifts to personal stories and ambitions. I order a double espresso and pour myself more sparkling water in anticipation of the long drive home.

"I'm curious, Susan. What gave you the vision to chase real-time data when the rest of us were still trying to get basic websites to load?" asks Ken, a member of Interval Research—the secretive Palo Alto think tank funded by Microsoft cofounder Paul Allen and known for hiring radical minds to prototype the future of media, interaction, and networked technology.

While sharing my origin story, I crack the sugar crust on a ramekin of crème brûlée with a delicate spoon. I recap my brief but momentous career and the exhilaration of creating something from nothing. The table responds musefully, and I absorb the warmth of connection and the great wines that long ago eradicated my initial nervousness. But as the mellow buzz wears off, I feel the return of my social discomfort in how to make a graceful exit.

After throwing business and credit cards on the table and instructing the waiter to divide the tab equally, we stand to say our goodbyes. As we part ways for the evening, we all exchange promises of future collaborations and friendships.

The night ends with Jonas's contagious optimism. He clasps both of my hands in his and says with conviction, "Susan, you've got the passion and vision that the Valley thrives on. We're all excited to see what you do next—and to help you get there if we can."

My smile lasts all the way home.

A marvelous destiny is unfolding, just along the horizon. I'm about to embark on a future based on the strong foundation I've created. What I have built has more than promise. It is certain to be a groundbreaking data revolution, and I'm telling the world—to their eager endorsement.

I had a bone-deep belief that if I didn't build something meaningful, I would disappear. It wasn't about ego—it was about survival. Being useful, impressive, or necessary felt like the only insurance against being ignored or discarded. Visibility wasn't just a career goal; it was a way to stay safe in a world that often had no room for people like me. If I could make something no one could ignore, maybe they'd stop trying to erase me.

Now I am on my way, and my troubles are all in the past. I'm young, healthy, and inspired by not only a dream but the ability to accomplish it. There's no sign yet that the dream might one day unravel.

CHAPTER ONE

THROUGH THE LOOKING GLASS

"You can't go back and change the beginning,
but you can start where you are and change the ending."

C. S. Lewis

1989—SAN JOSE, CALIFORNIA

My first bad decision didn't look like a mistake. It looked like a job offer. Sylvia is a New Yorker, an attorney, and hard as nails. She leads my interview at a massive optical and dental practice in a lonely office park close enough to the San Jose airport to smell jet fuel, but not close enough to feel like you were going anywhere.

I push down the familiar déjà vu of settling for less and power through. I am twenty-one, broke, and still convinced I can outsmart my gut. For the right paycheck, I can tolerate this fluorescent hellscape. I ask for fifteen dollars an hour.

What I thought was a bold step toward financial independence was really my first yes to the wrong life. I knew it was time to get a real, grown-up job, and this was the best offer I had at the moment. I quit my first job at a car wash when all my coworkers were going off to college or getting adult jobs, and it just wasn't as much fun as it had been when we were all in high school. I had never thought about a "career." At the car wash, I hadn't learned much, other than how to drink wine coolers and hang out with friends and drive over the hill to Santa Cruz for boozy bonfires.

I'd moved out of my childhood home for good three days after high school graduation when I was seventeen. College or a career were low on my list of priorities. One of the first things I did when I moved out of my parents' house was, stupidly, sign up for a Costco membership. I thought that since I needed essentials like milk and toilet paper, my paycheck would cover it. *How can I be broke? I still have checks!* That bumper sticker reality was my actual financial strategy.

My car wash friends lived at home, so their paychecks were for pocket money—buying clothes and going out to dinner. I didn't get it at first: that my paycheck wasn't *extra* money, it was survival money. I rented cheap rooms from random, often unsavory people or couch surfed when I mismanaged my funds and couldn't pay rent. Other than fleeting moments, I never felt safe or comfortable, but I knew how to be quiet, nearly invisible. When I decided to leave the car wash like all my friends had already, I pored over classified ads trying to figure out my next move.

I didn't have any skills besides cashiering and vacuuming filthy cars, but I found a job as a receptionist at an optical store in Los Gatos called Site for Sore Eyes. It did not take long to get bored of answering the phone, so our manager, a Japanese Mexican man with a beautiful accent who always dressed in stylish suits with a tie clip and pocket square, took me under his wing and became my first real mentor. He loaned me a textbook called *System for Ophthalmic Dispensing*, which was loaded with descriptions, illustrations, and photos teaching subjects like "effective diameter" and "measuring interpupillary distance." Each chapter ended with a quiz. I read the whole book and tried to copy my manager's style of speaking to customers. In doing so, I subconsciously, and hilariously, started to copy his Japanese Mexican accent for a while before I noticed it. I registered to take the biannual certification test, drove up to UC Berkeley, and passed on my first try. I suddenly had a career as a California licensed optician, and Site for Sore Eyes would need to hire a new receptionist.

I loved my manager, but the store's owner, Marty Dretch, was a sleazy guy who regularly parked his Jag in the handicapped space in front of the store—with no placard and no shame, even when there were empty spaces right next to it. He divorced his wife and co-owner of the company to date a receptionist at one of his other stores—a girl my age named Chessica. When he made her the new manager of our store, I knew right then I had to quit.

Now, sitting across from Sylvia, I feel mature with a full year of opticianry experience under my belt. Sylvia wants to give me nine dollars. I make her a deal.

"I will work here for a month for nine dollars an hour," I say. "In thirty days, if I have not knocked your socks off, achieved the highest sales numbers you've ever seen, and delighted customers with my knowledge, professionalism, and charming attitude, then you can fire me on the spot, and I will walk away. But if I have reached those standards, in your opinion alone, you will sign me for fifteen dollars per hour, plus commission."

Sensing no pushback, I add that I also want performance reviews every six months and an opportunity for a raise once a year.

After a beat, Sylvia thrusts her hand forward.

"Welcome to Skyport Dental and Optical Group."

I can see that I have impressed her.

I do have a couple of obstacles in my way. One of them is Wende Rose. She's even younger than I am and doesn't have her optician's license, but she does have the keys to the file cabinet—and a confident personality. Wende has been placing orders for optical supplies, frames, and contact lenses for about eight months already, and she is holding fast to that seniority. No one sits in Wende's chair. She stays in the little alcove with the computer and phone, placing orders and calling clients for appointments and fittings. I see Wende as a threat at first, but then I recognize myself in her. Even at our tender ages, we develop a mutual respect for each other's work ethic.

Ann Dixon is hired shortly after me. She's never had a job in the optical business before. Ann is sweet and gracious with a beauty that shines from within. She drives a cute little blue Nissan sedan with a personalized license plate professing her love for the rock band INXS. Soon we are joined by Tammy, a *Little House on the Prairie* character all grown up.

I receive the raise I asked for. Sylvia doesn't even wait for me to bring it up. After exactly one month on the job, she has the accounting department increase my hourly wage to fifteen dollars. Despite the owner's investment in the staff, business doesn't take off at Skyport Optical. On a typical day, employees outnumber clients six to one. That leaves a lot of time for my office friendships to bloom.

One night, we girls all go out to one of Ann's haunts, the Oasis nightclub, and Tammy turns out to have an alter ego. In her frontier-style blouse and long khaki skirt, which stand out in the edgy, alternative rock nightclub, she gets super drunk, acts like a raunchy bitch, and fights with everyone. Then she brings a random guy home to my apartment and has sex with him on my white couch while she's on her period. She never lives that down with the girls in the office, and I permanently borrow a throw blanket from my mother's house to cover the stain on my sofa.

The optometrist in our office is a disheveled alcoholic who clearly has the wrong prescription in his own glasses. He has a tiny office in the building we share with the larger dental practice and surgery center. Our doctor comes in late almost every day and doesn't even hide the bottle of cheap vodka in his office, where he is often found "sleeping" in the exam chair. We take turns waking him before appointments while another one of us distracts the patient with our collection of frames.

It is here at Skyport Dental and Optical Group where I meet my future husband, Cliff.

Cliff Mulford works in accounting, and he brings the mail to our side of the building every day. The optical office is overstaffed in case of a busy rush that never happens, so we spend most of our time polishing glasses, organizing and reorganizing the displays, dusting, and gossiping. When Cliff comes in, we women have very little work to do.

Cliff is quiet and has a sloppy gait. He's slightly overweight, but I think that's maybe just how he's built. Cliff smokes cigarettes like Ann and I do, so we take breaks together outside and enjoy making conversation with the only guy our age in the office.

My first impression is that Cliff seems dull and harmless. A guy from Santa Cruz who doesn't surf. I fell for the bad boys at the car wash and surfers before that. They never had their own cars and were always borrowing mine—a nuisance—so I am interested in trying something new. Cliff has a white Volkswagen Jetta. I decide to date him.

There are no fireworks, no spark, but he's a reliable, pleasant guy who seems like a more sensible choice than the young surfers, musicians, and other charming disasters I attract. And I'm burnt out from being alone in the world.

I know I'm getting older, and I still haven't had my first real boyfriend. I feel the pressure of aging even though I've just entered my twenties. Without any passion for a specific career track, no hobbies or life experience, and zero interest in higher education, the only thing I know for sure is that I eventually want to have a family. When I look at my daily routine, it feels like I'm going nowhere, and I have the palpable sense that I'm just taking up space and contributing nothing to the world. I'm bored with my studio apartment in a shady section of town and bored with life altogether. So, in practically no time, Cliff and I get a loft in downtown San Jose and start playing house.

It doesn't take long to suspect I've made a grave error.

A month after we move in, I take my annual preplanned trip to Hawaii to visit my childhood friend Greg and the group of surfers that pile into the dorms each summer at the University of Hawaii. On the islands and away from the Bay Area, I suddenly feel full of possibility and opportunity. For two weeks every summer, my budget vacation consists of sleeping on a cot in the dorms for free, crawling into the back of cruiser wagons with surfboards on top, and spending each day lying in the sun on the beach or on friends' boats. I'm not a student there, but I party along with kids on the cusp of adulthood whose parents pay for their education and room and board, steering them toward their future. Living in close proximity to students who were dreaming big dreams was contagious, and I wanted to join them in this lifestyle. I speak with Ann about starting our own contact lens distribution business.

"Tourists lose lenses in the water," I tell her. "They need quick replacements. We could set up a little kiosk just off the beach in Waikiki, drop flyers at the hotels, and offer same-day fittings. We would call their doctors, get the prescriptions." Ann just got her contact lens fitter license, and I can easily renew mine. I know it is a solid plan.

In the glow of my first real business idea and the great friendships that surround me, I feel excited and energized like never before. I realize that Cliff is just a placeholder—a cop-out because I don't know what to do with my life. The physical distance between us starkly reveals that I'm imagining something in him that doesn't exist. Does it make sense to say that I chose him simply because he looked like a father I had seen, an uncle of a friend at a backyard barbecue, who was playing on the grass with all the little kids. I'd thought, *That's what I want. A man who will be a great dad.* I remember that scene and the way the dad looked, a lot like Cliff— pudgy, no fancy clothes, playing and spinning the kids, throwing them up in the air. I like the potential of Cliff, the stability I've imagined in him, but I soon learn he's content to stay still, and I'm not done moving forward.

Before this trip, I was bored and lonely, uninspired and sick of waiting for my life to begin. I felt no real passion for anything, but I was at least aware of it. Knowing something needed to change sparked me to move in with Cliff. But now, I am beginning to understand that something more is possible, and I recognize that getting a live-in boyfriend is not the solution to my apathy.

When I return from that trip to Oahu, I'm resolved to break up with Cliff right away. I'm kind but firm about it. He seems to take it well at first and doesn't argue the point. I relax, grateful that the breakup was so much easier than I thought it might be.

But, after a few tears, Cliff spins out emotionally. His sadness turns into something I've never seen from him before. First, he's angry, then suicidal.

"You go on vacation with your friends and come back with all these new ideas, and you have the audacity to *dump me?*" he bellows. "We signed a lease on the apartment!"

I don't know how I'll do it, but I mumble that I'll pay him back his half of the deposit for breaking the lease. I'm ashamed that I'm backing out of my rental commitment, but I know I do not want to be with this man.

All the while, I remain calm and offer him no fuel. As the last of his anger fades, his demeanor shifts once more. He walks out alone onto our balcony and begins to cry audibly, his wails intensifying with each breath. As the sun begins to set in the distance, I hear him talking to himself.

"You know that I could jump," he says to no one but loud enough for me to hear through the sliding glass door.

Guilt strikes me, and I inch toward the balcony. Did I hear that right?

"Are you going to be okay?" I venture softly.

He tells me that he wants to die, then drops a startling footnote: He has been diagnosed with chronic depression.

"I'm supposed to be taking Prozac," he sniffs, "for my depression. But I stopped." His face is red and puffy. "I feel so out of control."

His wailing intensifies again, large tears plummeting down his cheeks. I suspect that I'm being manipulated, but I decide to believe him—just in case he really is thinking of hurting himself. I have no experience with depression.

Instead of walking out, I tell him I will stay for a little longer. Postponing my own needs and putting my wishes second to his gives me a feeling of empowerment. Because I'm young and inexperienced, that's what I think empowerment might feel like. I decide I'll leave in a couple of weeks. By then, he'll be better able to handle the breakup. This is the kind and responsible thing to do, isn't it? Even though we have only been dating for a few months, I will give him a chance to absorb this change.

As time creeps by, I transform into one of those frogs slowly boiling to death in a pot of water. The incrementally rising temperature masks the mortal danger I'm in. Two weeks turns into two months, and all the while, Cliff is agreeable and easy to get along with. I forget my determination to break up with him. I forget the longing for growth that Hawaii reignited in me. The little dream Ann and I concocted to start our own contact lens delivery service to hotel guests in Waikiki now feels more and more like a silly vacation fantasy.

And then, before I know it, I get pregnant, and I have to reassess everything.

I've come far this year, and now I'm heading toward a major milestone I'm not quite ready for. I've learned basic life lessons like giving up my Costco membership and instead going to the bargain market to buy one half gallon of milk or one frozen burrito at a time. Through trial and error, I figured out how to survive on my own, and now I will figure out how to survive with my baby.

I think about the young woman I was when the year began, the one who met Sylvia and confidently negotiated her first grown-up job, finally sculpting a bit of a future. I remember the first spark of an idea: to start a business, to live freely in the Hawaiian Islands with friends, to pursue only what was mine. Then came the sharp turns I couldn't have predicted. Now, with this tiny child growing inside me, everything must shift again. The pregnancy anchors me and unmoors me all at once. My dream of a family is arriving, not on my terms, but undeniably real. I'm determined to make it work. Maybe it doesn't matter that it's too soon, or that I'm caught in a life I didn't quite choose. The future stretches wide and full in front of me, and I know I exist—somewhere, somehow—inside it.

CHAPTER TWO

CRACKS IN THE FOUNDATION

"Our lives are shaped as much by those who leave us as by those who stay."

Anonymous

"**M**ochi! Let's play!" I call out to Moch, and she comes bounding up to me on the floor of the family room in our one-story ranch-style house on Lois Avenue, a neighborhood where daddies go to work and mommies stay home with their babies. Mochi is a special dog. She has eyes as rich as chocolate buttons and fur as soft as whipped cream frosting. I cherish her. Mommy and Daddy got her when I was still inside Mommy's tummy. My uncle Fuji, Daddy's best friend, is Japanese, and he gave Mochi her name. He told me it means "little tea biscuit" in Japanese, and I think it's the best name ever. I love her so much, like she's my twin sister, but I know she isn't. Sometimes, I feel a little jealous because Uncle Fuji knew Mochi even before I did!

Uncle Fuji is a vibrant presence in my life, always bringing warmth and joy whenever he's around. Though he's not my actual uncle, I call him that because of the close bond between our families. Compact and strong, he moves with a lively energy, always ready for the next playful moment, whether it's making funny faces at me or pretending to be a kung fu master in a mock battle with my finger puppets.

But when he and my daddy get to talking about philosophy and big ideas, there is nothing silly about it. They are thoughtful, intelligent men of different backgrounds who enjoy speaking on high-level subject matters. I listen to them with my paper crown atop my head and my blanket wrapped around me like a princess's cape or while hiding behind the sofa and playing with my doll Mrs. Beasley. In my finger puppet shows, I pretend my puppets talk like Daddy and Uncle Fuji do. When I interrupt the men, they never send me away or talk down to me. Instead, they talk to me like I'm special, which builds my expectations for the world ahead. I am in kindergarten now, and they treat me as important as I feel.

Our house has a big front yard and an even bigger backyard. It's one of two houses we own on this street in Sunnyvale. My aunt Caroline and uncle Joe live in our other house down the street with my cousin Penny, who is the same age as my brother Evan, and her brother, Sammy, who is a baby. My grandma and her sisters, all the aunts and uncles, the cousins— everybody is always in the backyard at our house or Aunt Mary Ann's house, usually barbecuing hot dogs and eating fruit salad and drinking soda pop almost every weekend. Daddy keeps the yard spick and span for the whole family.

And then one day, I feel something bad happening. Grandma is at our house, and I see her with her arm around Mommy, trying to make her feel better. I don't understand what's going on, but my tummy feels yucky. I want to see everything I can't see and understand everything I can't understand. I creep toward the kitchen to listen in.

"Be real quiet, Mochi," I whisper to the pup, who follows at my heels. Mochi looks up at me in understanding.

Mommy is crying.

Grandma Emeline works at Lockheed, next door to NASA. Grandma has a very important job. She's always busy with work, so her being here now tells me something is wrong. Mommy sees me and dries her tears on her apron.

Grandma gives Evan and me chocolate milkshakes from the blender. This is extra special because it isn't anyone's birthday. Milkshakes and Grandma on a school night? That's a treat, but something feels bad about it—and I'm not sure what.

Mommy sits down in the sunken living room, and with a serious voice, she beckons for us to come in. Grandma reaches to help us with the milkshakes that we struggle to hold in our small hands, but Mommy says we can bring them in.

Wow! How did we suddenly get so grown-up?

"I don't think Evan can have a milkshake on the carpet—" I start, but Mommy shushes me.

"It's okay, Susie. Just this once," she says, tired but firm.

I'm just trying to remind her of the rules. Evan is probably going to spill his milkshake. She shouldn't let him have one in here.

Oh, well. I shoo away the thought and focus on my own milkshake. Evan and I sit, sipping, on either side of Mommy on the couch.

Just then, my tall, handsome daddy comes home, a bit late from work —a new habit. He works in a real estate office to help all kinds of people buy houses just like ours for their families. He's dressed in his usual black business suit, white shirt, and skinny tie. He flashes his biggest full-face smile at Evan and me, puts down his briefcase, and kneels low on one knee with his long arms opened wide for hugs, just as he does every night when he comes home from work.

I set my shake on the coffee table and run to him, but Evan spills his. Grandma, who's standing nearby, crouches down to dab the peach-colored carpet with a wet cloth.

Mommy interrupts with tight, unfamiliar lips. "Daddy is here to say goodbye," she announces stiffly, as if doling out a punishment.

I look up at Daddy with big, round eyes and wonder what's happening. Goodbye? Where's Daddy going? He just got here.

My heart races, and I wrap my arms around Mochi for support. After Daddy hugs us again, he gathers the suitcases waiting for him by the door and, looking so sad, walks out.

1975—CAPITOLA, CALIFORNIA

Daddy lives with Claire now, and she's our new stepmom. Before Daddy and Claire have their baby, I spend weekends and part of summer at their beach house in Capitola, where they teach me how to dig for wild clams. I watch Claire in her bathing suit, striking and tall with her bouncy dark hair. She screws her feet into the sand in shallow water about ankle deep, rotating her hips from side to side, and tells me to copy her. I turn my body, screwing my bare feet into the wet sand until I feel something like a rock with an edge. It's a clam! I quickly bend down and push my hand into the sand next to my foot like Claire does. She pulls a clam out. Mine gets away, but I try again until I get the hang of it. When I do, I scream in delight. I place my beautiful, plump clam in our bucket and follow along again and again until the bucket is heavy. Daddy carries it up the small hill toward our house.

At home, he fills the bucket of clams with seawater and cornmeal and leaves them in the garage overnight. Daddy explains that the clams eat the cornmeal, which cleans out their intestinal tracts, and they spit out all the sand they've consumed. Then, we cook them in a covered skillet with garlic, butter, and white wine and serve them with a crispy sourdough baguette and fresh steamed artichokes for dinner.

Wild steamers is our favorite recipe in summer, a perfect blend of precious moments on the beach and our harvest from the sea. Always the teacher, Daddy winks at me where I stand tiptoe on the stool next to him in the kitchen as he plates each delicious spoonful.

I love going to the beach with the plastic pail I use for building sandcastles. I now have a new hobby—getting dinner! I walk down to the shore all by myself and dig for clams and bring them home in my little beach pail for our family to eat. I feel wonderful doing this work. I'm complete, capable. I feel more grown-up, after all, I am in the first grade now. I can think of no wish, desire, or need that isn't met by clamming at the beach. I have everything I need, including the knowledge of how to get food on my own so I can make a delicious meal for us. I'm content and aware of a new kind of satisfaction. I know something powerful now: how to feed people. It feels like my own precious talent. I don't know any other kids who can do this.

When they can't get a babysitter, Daddy and Claire bring Evan and me to fancy bars. I love the warm leather booths and the enticing spectrum of colorful sugar layered in glass jars on the tables we sit at while our fabulously dressed-up parents drink martinis with olives or tiny white onions. There are always large groups of friends laughing and talking and listening to a pianist in a tuxedo pound away at a shiny black grand piano. Evan and I fall asleep beneath the table, warm and comfortable on the leather booths.

Even at home, we know how to behave. Evan and I are very good kids. Daddy and Claire always have new coloring books for us, and Claire often colors one side of the page while I color the other. I sharpen our crayons, peeling some of the paper and inserting the blunt ends into a hole in the back of the crayon box until I make a smooth point. Adults always tell me that they are impressed with my art skills. I keep Evan busy with books until he falls asleep since he is littler than I am.

Sometimes I get annoyed at Evan when he won't sit still, but that's just because he can't stop wiggling. One of the grown-ups takes him 'til he settles down. He is never bad; he just can't sit in one spot. We never get in trouble for anything with my dad. Daddy takes the time to talk to us, and he's really nice. My mom doesn't really talk to us. She is always sad, tired, or annoyed. She never looks at us in our eyes like the other grown-ups do. Well, she doesn't look at me, anyway. Maybe she does look at Evan. Mom always says Dad favors me, so she has to treat Evan better than me to balance things out.

Plus, Mommy always takes extra care of Evan because she says I am fine on my own. "One thing about Susan," she says, "you never have to wonder what she's thinking."

I'm not exactly sure what this means, but I guess it's a good thing.

"One thing's for sure," she says, "we don't have to worry about Susan."

My mom is saving up for Disneyland. We have a big glass jar of coins, which I tally up as if it were a jelly bean jar at the school carnival. Guess how many jelly beans are in the jar and you get a prize. I know at all times how much money is in that jar because I stack up the quarters and dimes to count them. Mom says when we have enough money, we'll take a vacation to Disneyland. That means that when we're out shopping, Evan and I can never ask for anything.

"The money has to go in the Disneyland jar," Mom says.

She tells us how wonderful Disneyland is. Of course, we know from TV that Disney is the happiest place on Earth!

It's around this time that my aunt Brenda and uncle Jack take us to my very first movie: *Mary Poppins* at the drive-in. My cousins, Evan, and I all cuddle up in sleeping bags in the car in the parking lot. I love Mary Poppins and Bert and the kids and the singing and the way they talk like Claire, who has the same British accent as Mary Poppins. If that is what Disneyland is like, I'll sacrifice anything to get there! Yes, I can deny all my impulses to save money for this trip, and it's a goal I vow to take seriously.

Mom is dating through a group called Parents Without Partners. The name seems corny, a word I learned in my first-grade class from my best friend, Lisa Reed. To Lisa, everything is corny, including the rich girl's purple Donny Osmond lunch pail that we secretly covet.

My mom is sad all the time. She takes classes at nursing school during the day and at night she works the swing shift on the factory line at Fairchild, where they make tiny parts for computers. It seems like her third job is finding a man so that she can finally stop working and stay home again.

After the divorce, we move into Grandma's apartment about five miles away in a much more run-down, industrial area with a busy street, empty lots, and saggy apartment buildings. Grandma moves into one of the apartments upstairs. Over the brick fireplace in the dark fourplex apartment building, Mom hangs a picture she painted of Emmett Kelly's sad clown character. There is always a dampness, a clean-but-never-fresh smell to the apartment. We have thick green shag carpet that holds tight to the sweaty canine smell of my grandmother's tramp, old Rags.

Since Mochi lives with Daddy and Claire now, Daddy gives me a rescue dog for my birthday. A real basset hound, just like I wanted. I name him Bernie. He howls and sits around looking cute. After two weeks, my mother gets fed up with the lazy old dog that howls all day when we're not home and drops Bernie off at the pound while I'm at school. I cry and cry and cry, because I didn't even get to say goodbye to him.

Mom starts finding matches right away in Parents Without Partners. A nice man named Tom is probably the best of them. He makes pizza from scratch in his sun-filled kitchen, which we love, but his kids are pretty crazy. They're fun but out of control, meaning now as I look back, they just acted like kids. I, a mini adult, understand when my mother tells me she can't tolerate them. Too much work. All the same, I can't help but love them. They are so free! Tom's daughter especially reminds me of Peppermint Patty from *Peanuts*. My mom says I'm more of a Lucy, always playing tricks on people.

"You force people to like you," she says. Then she follows up with a compliment telling me I'm too smart for my own good.

Evan has a favorite blue blanket, making him a natural Linus. But I think Mom is wrong about me. I'm Snoopy, of course! I think I wanted a basset hound because I got beagle and basset hound mixed up.

After Tom comes a tall disco king named Bro. Seriously, that is his name. Bro. He has a groovy seventies-style pad and a waterbed, which means his place is probably a condo. When I say he's tall, I mean really tall! Almost seven feet. He has a bushy mustache that I can see if I crane my neck. I think Mom likes him, and he is nice enough. His kids look like young surfers, but they're with their mom most of the time. I don't think my mom likes that Bro has other girlfriends, though. She wants to be the only one.

Soon enough, some of my mom's friends set her up on a blind date. Salvatore Caputo Jr. is not from Parents Without Partners, and he doesn't have kids. When it's time for us to meet him, he takes us to the Ringling Bros. and Barnum & Bailey Circus. Evan and I have never seen anything like it. Flashing lights and toys are beckoning to us from everywhere, even draped on guys dressed as clowns. If there were ever a situation tailor-made to get children to beg for toys, this is it—and we do. We are treated to too much cotton candy, and all that sugar gives us the insane and thrilling confidence to ask for anything and everything we see in the dazzling glow that cuts through the dark circus auditorium. Spinning, fluorescent tops and wands with cascading neon rainbows dangle in our faces, and we join the throng of kids going wild for it all.

Our unbridled mischief prompts Mom to pull us back to the car early, where she scolds us. By now, we're crashed out on sugar like slugs in the back seat. I want to melt into a hole when Mom's date, Mr. Caputo, tells us how badly we've behaved as he begins to drive us all home.

When he's done with his verbal thrashing, he looks over at Mom in the front seat and takes her by the hand. "Don't worry," he says. "I'll get these rugrats under control."

Mom smiles back at him like he is some kind of hero.

CHAPTER THREE

THE BALLOON EFFECT

*"The most common way people give up their power
is by thinking they don't have any."*

Alice Walker

1990—SAN JOSE, CALIFORNIA

When Cliff and I share the news of my pregnancy with my mother and stepfather, Sal makes his opinion plain.

"You get married, or you have an abortion," he says, then walks out of the room.

My mother does most of the wedding planning because I'm too sick and too weak to make any decisions—but also because she's genuinely excited. I'm her only daughter, and now that I'm getting married, I'm suddenly a priority. I feel like her favorite, for once. She chooses burgundy ink on cream-colored paper for the invitations—not because either of us likes the colors but because she says they are appropriate for a late autumn wedding. I don't have the energy to suggest aquamarine on white, which is what I really want. I'm fairly happy to be pregnant, getting married, and receiving lots of attention from my mother, who enjoys planning the happy occasion.

No one ever questions whether Cliff and I are in love. We are not, but that doesn't seem to be important to anyone, including us.

My mother instructs Cliff and me to make a list of people we want to invite. She's paying for the wedding and insists our reception be held at the club where Sal is a member, which boasts a historic, Gatsby-esque ballroom. This will be an opportunity for them to show off their wealth to Sal's colleagues from the Rotary Club. Because it's short notice, though, its only availability is the Saturday of Thanksgiving weekend. The club has another event that evening, so our wedding is scheduled for 10:00 a.m., with the reception immediately following. We need to be out of there by 3:00.

For the first four months of my pregnancy, I'm terribly sick. I'm so weak and vomiting so much, I'm forced to quit my job at Skyport Optical. I enroll in a vocational program at a high school near my mother's house in Almaden Valley to learn the basics of court reporting. I can't keep anything down, and I throw up several times per day, so this flexible program that is paced for students to excel at their own speed is perfect. I'm already very skinny, but I lose another fifteen pounds. At five foot five, I weigh only ninety-five pounds on my wedding day, three months pregnant.

I wake up on the day of my wedding at my mom and Sal's mansion, which, on a street named after them, backs up to the Santa Cruz mountains and overlooks Silicon Valley. My mother has her hairdresser come to the house to style me and apply my makeup. She gives me a French twist, just like I want, but she teases my bangs and lacquers them in Aqua Net hairspray so that they stand straight up over the headband of my veil. In the bathroom, still woozy from my latest round of vomiting, I try to push my bangs down, but they pop right back up. I don't have the energy to try harder or ask the stylist to fix it.

It's my wedding day. When I look in the mirror, my skin looks green and my eyes hollow. I wear a beautiful yet borrowed antique satin gown. The whole thing feels rushed and lacks sincerity. I am not the image of the perfect bride I'd hoped to be.

My parents have invited everyone they know for a sit-down wedding dinner during the lunch hour, with a choice of Dijon chicken or sole meunière. Cliff has about ten friends in attendance with their plus-ones, and I have a few more than that. Midway through the reception, Cliff's friends complain that the open bar doesn't cover beer, only wine from Sal's winery, and I need to fix it. They're a bunch of twenty-two-year-olds from Santa Cruz. They came to the wedding for the beer—that's part of the deal! What a disaster. We don't drink wine! I can't believe that's what my parents call an open bar.

When it's time to cut the cake, we intend to follow tradition in which the bride and groom feed each other a piece. Cliff goes first, and he smashes the cake on my face, in my hair. The satin dress, a fifty-three-year-old heirloom borrowed from Sal's mother, is covered in frosting. She watches in horror as her dress is defiled. Cliff grabs me, plants a clumsy kiss on my face, laughs, and holds his fist up to his friends like a conqueror. I'm humiliated and smeared with frosting.

My mother takes me upstairs to help me change into my going-away outfit a little early. While I wipe cake from my face in the mirror, the party is rocking downstairs, and people are dancing. As I take off the soiled satin dress, I realize I left my street dress hanging in the back seat of my mom's Lexus. She goes down to get it for me—and never returns. I wait in the private bathroom upstairs and cry, trying not to ruin my makeup.

After a long twenty minutes, I venture into the hallway, half naked, the satin wedding dress draped around me for cover.

"Mom!" I call entreatingly down the stairs.

I catch sight of her shaking hands and saying goodbye to everyone who decided to leave after the cutting of the cake. I understand that this is *her* party, not mine. No one has even noticed I'm gone.

Cliff's mother gifts us three nights at the Highlands Inn in Carmel-by-the-Sea for our honeymoon. When we arrive there in the dark the night of the wedding, it's raining. I sit on the bed, thoroughly exhausted and nauseous, and Cliff asks the bell boy to build us a fire. When we are alone, Cliff puffs up his chest, ready to make an announcement.

"Now that you're my wife, everything changes. Starting now," he says.

I listen, numb, as he recites a set of rules he expects me to obey. Going forward, I will perform specific sex acts several times a day, starting tonight. Romance, love, pleasure, communication—none of these were mentioned.

Even worse, I am still vomiting nonstop due to debilitating morning sickness. But my mother's marriage advice is ringing in my ears: "There is only one rule to a happy marriage. Never say no to sex. If you do, he will find it somewhere else, and you will have no one to blame but yourself." When she spoke these words to me, I was sad for her, but now I am facing that very dilemma.

Back at our loft apartment in San Jose, I try to make the best of it, but when my husband orders me to perform sex, I negotiate hand jobs so that I can be involved as little as possible. I keep tissues next to the bottle of Jergens lotion by the bed so I can discreetly spit out his semen if I have to give him a blow job when hand jobs become too obvious a distancing technique. I do everything I can to avoid actual intercourse with him.

"What's wrong with you!" he berates me when demanding sex, "You're *frigid!*" This becomes a familiar line. Poor Cliff. He married a frigid woman.

Cliff has several tricks up his sleeve to lure me into submission, like withholding grocery or diaper money until I perform. Sometimes he empties the gas tank and refuses to refill it. He offers me a dollar or, when he's feeling generous, five dollars per hand job or blow job when he knows I need to buy food. I am far too embarrassed to tell anyone about this. I haven't even considered what would happen if I tried to leave.

I know this is an unacceptable situation, but I also feel trapped. My mother and Sal have just paid an extraordinary amount of money, ten thousand dollars, for a lovely wedding reception. Then I came home to all these beautiful wedding gifts: a full set of china in the pattern I had chosen, Oneida sterling silver-plated flatware, Waterford crystal candle holders and bowls, a Lladró statue, a KitchenAid stand mixer, and a full set of All-Clad pots and pans. Plus, there's the money my parents' friends gave us. I am pregnant, unemployed, physically frail, and suffering Cliff's subjugation. Fleeing this marriage is impossible in my condition. I have to suck it up.

With the baby coming and my constant queasiness limiting my ability to work, Cliff's mother, Dorothy, offers us one of her rental homes in Capitola and discounts our rent. Eventually, she says, we'll be able to buy this house from her at a low cost. While I'm grateful for all the gifts and excited by the idea of living in Santa Cruz near one of my favorite beaches, I can't face the truth that despite appearances, my predicament is a mind-numbing, horrifying ordeal. Instead of taking walks on the beach, I become agoraphobic, afraid to open the curtains or go to the mailbox for fear that the neighbors will engage me in conversation and I'll have to throw up in the street gutter. When Cliff is at work over the hill in San Jose, I keep the curtains drawn and pretend not to be home.

When Emma is born in June of 1991 at Kaiser hospital in Santa Clara, I am twenty-three years old and have been married to Cliff for just under seven months. I wanted a natural birth with "rooming in," and I attended all the Lamaze classes at the hospital with Cliff, learning as much as I could about what to expect. I even wrote out a proper birth plan, but no one even looks at it once I'm in labor. Since I'm three weeks overdue, my doctor decides it's time to induce.

After eighteen hours on Pitocin, an intravenous drip of artificial oxytocin used to induce labor, I begin passing out between contractions. My cervix stops dilating, so a masked doctor appears and, without speaking to me, orders an epidural by injection of lidocaine or morphine into my spinal cord. When the anesthetics kick in, I am able to relax. Over the next few hours, nurses pop in occasionally to screw probes into the baby's head, which is still inside me. When they reach in with gloved hands and fumble to screw in the little probes that will monitor her heartbeat and vitals through her skull, the pain is excruciating, but I am worried about the baby more than myself. I did not hear of this procedure in Lamaze class or read about it in any books. Nothing is going as I expected.

My mom, Cliff, and Cliff's mom, Dorothy, are in the delivery room with me, assisting the nurse. All my aunts and uncles and my grandma Emeline have been rotating in and out of the waiting room, my uncles slipping away to the bar down the street and my aunts popping in during their lunch breaks from work. It's a busy night in the maternity unit, with fifteen other babies being born at the same time. The hospital is severely short-staffed and doesn't have enough delivery rooms for all the expecting moms. Our nurse is alone and absorbed with filling out papers in a massive binder. When I become coherent enough to speak again after the epidural medication kicks in, I ask what she's doing.

"Procedure requires me to document everything that happens during your birthing experience," she says, still scribbling.

The whole binder is just for me. I'm shocked by this dysfunctional system. Instead of providing medical care, the nurse is doing paperwork while my mother and mother-in-law hold my hands and monitor the vital signs of the baby and me through my contractions.

After twenty-four hours of labor, night falls, and the masked doctor again materializes. Dorothy and my mother support me, lifting my body and pushing with me.

And then, in all the commotion, like an angel calm and pure, my baby arrives.

I take her in my arms, and she fits perfectly. In an instant, I feel an incredible euphoria, a transformative love that makes me forget all the pain, all the confusion. My sight becomes clear, and, looking at her, I realize that everything else in my life so far pales in comparison to this moment. Nothing has prepared me for this kind of love. The past is mere shadow. This is pure light.

I look at her scrunched-up face and squinting eyes, wishing the hospital lights were not so bright. At noticing the sores on top of her head from the probes, I feel incredible regret that she had to go through that while still in the womb. Then I remember the advice I read in every pregnancy book about the importance of colostrum in providing the essential nutrients for long-term health, and I give her my breast to comfort and nurse her.

Once my mom has made sure the baby is perfect, she goes to the waiting room, where my family is camped out.

"Emeline Lucia is here!" she exclaims.

The room erupts in joy and cheers. I am my grandmother Emeline's first grandchild, and this baby is her first great-grandchild. I've named baby Emeline—Emma for short—for my grandma, adding the ethereal "Lucia," meaning light.

Holding my child for the first time, I feel my life transform at once. Suddenly, everything about the way I see the world, and myself in it, shifts in a seismic realignment back to what I knew when I was a very young child—my natural confidence and undeniable belief in a future full of love, adventure, and possibility. A hidden compartment reveals my true self, a knowing that has been buried and waiting. Now, my intuitive knowing floats back up to the surface from the dark depths of my adolescence. Inexplicably, I feel a strong desire to own land and grow food and gather eggs. This dream appears out of nowhere, or from some forgotten place deep within me. I had been content to live in random apartments as a young single woman—perhaps my lack of motivation was because I didn't value myself. But the minute Emma is born, I value her and then myself just as well. Her birth reignites the desire for the life I knew I was meant for from my early childhood.

Becoming a mother somehow unlocks my treasure trove of self-trust, my strength, and the courage to choose my intuition over the outside voices so confidently commanding me to behave, comply. Because of this child, I now know on a cellular level that there is no going back to ignoring the truths that have always lived inside me, in my heart, in my body, and in my mind.

Twelve hours later, that knowing intuition will save the child who gifted me with it.

NEW YEAR'S EVE, 1974—SUNNYVALE, CALIFORNIA

Things with my mommy and Mr. Caputo are getting more serious, and he starts attending family functions. The idea is that he will become my stepfather someday, but I really, really hope Mommy changes her mind about him. He's not a good person at all! He just doesn't fit with my family. He is nothing like my real daddy: nice, kind, open, generous. Mr. Caputo uses curse words in practically every sentence. He's constantly insisting that we need to learn *respect*, that we must obey his every word.

On New Year's Eve, there's a party at Aunt Mary Ann's house. She decorates to the nines. I help her with the shish kebabs, cheese balls, ambrosia, and other hors d'oeuvres. Then I dash back and forth from the kitchen to the living room, delicately arranging crystal bowls of mixed nuts on the coffee tables. I have a green balloon tied to my wrist with a ribbon and a little cone-shaped party hat. Everything is perfect—except that Mr. Caputo is here. He wants us to call him Sal now. I can tell Aunt Mary Ann doesn't like him either. They've argued before, and tonight she isn't even speaking to him despite hosting him in her home. Maybe my mom and Aunt Mary Ann had a fight, too, because I don't spend afternoons at her house baking and playing dress-up and talking about the future anymore.

Evan and my little cousins are playing hide-and-seek, but I'm content to sample the food and play with my green balloon until Aunt Mary Ann needs me again. But Sal has a different idea.

He's a heavy smoker. My mom smokes too—and so does Aunt Mary Ann—so his smoking isn't anything new to me. Tonight, though, he lights a cigarette and calls me over. I know by now that I'm supposed to do whatever he says, so I pad reluctantly across the living room, my green balloon trailing behind me. I instinctively stand just out of his reach.

Sal takes a long drag from his cigarette, and then, before the smoke has even escaped his pale lips, he thrusts the burning end of his cigarette at my gently swaying balloon. He is trying to pop it!

"No!" I cry, stepping back.

Sal cackles. He misses the balloon, but it doesn't matter. The game has only just begun. He takes a sip of his drink, then chases me across the living room, jabbing his cigarette wildly through the air.

I bob and weave through the crowd, narrowly avoiding him. "No!" I cry again, and his manic laughter finally withers. I can see by the change in his face that my defiance has started to anger him.

At last, he corners me, his cigarette menacingly close to my face. I hold the balloon tightly in my arms, trying to keep it safe. But Sal can't let it go. He has to show me who's boss. He grabs me and tries to wrestle the balloon away from me. He's laughing again, but there's nothing funny about it. It's a mean and awful game he's playing.

I'm horrified. Why is everyone letting this happen? This mean, ugly, stupid man wants to pop a little girl's balloon with a cigarette. He really thinks it's fun. How disgusting!

Eventually, Sal mashes his cigarette against the tight skin of the balloon still cradled in my arms. Though I knew it was coming, the shock of the pop is too much for me, and I start to cry. I cry, and he laughs hysterically, a winner.

"Don't be such a baby," he says when he catches his breath. "It's just a game."

As my tears turn to anger, his laughter turns into annoyance.

"Oh, come on," he spits. "It's not that bad. It's just a balloon. Go get another one if you love it so much."

I don't want another balloon. It's not about the balloon. Sal goes back to his drink, and I'm alone again with the latex carcass and plastic string.

In a photo that my mom arranges neatly under plastic film in our family album for all to see, I'm cowering in the corner in my yellow dress and party hat. Sal is in the foreground, a horrid laugh painted across his face. He's ashing a cigarette into a glass tray. I'm glaring at him, clinging desperately to my balloon. You could almost hear it in the click of the camera—a door to my childhood closing, closing, shut.

CHAPTER FOUR

HOUSE OF MIRRORS

"The world breaks everyone, and afterward,
some are strong at the broken places."

Ernest Hemingway

1991—SANTA CLARA, CALIFORNIA

Every expectant mother imagines the day she welcomes her child into the world. The narrative I spun for myself—the natural labor, the attentive staff—turns out to be miles away from reality, but the arrival of my beautiful baby girl erases any pang caused by that unfortunate hospital experience. She is everything I dreamt of and more. I can't imagine ever being parted from her. Or that anyone would ever try to take her from me.

It must be almost midnight by now. Still, I object when a nurse comes to take Emma to the nursery. I have a strong preference for "rooming in," a hospital practice that allows a newborn to stay in the room with her mother rather than being carted off to the nursery. My mother painted a picture of this for me when my little brothers Giovanni and Matteo were born—how she opened a drawer next to her hospital bed and the baby was there whenever she wanted him. This approach encourages bonding, promotes breastfeeding, and allows new parents to get accustomed to caring for their baby in a supportive environment, and I look forward to experiencing it for myself. As such, I expect my baby will be with me all throughout our hospital stay unless I need a break. In the birthing room, I'm still in my groggy haze from labor when I realize I'm receiving a shot in my thigh.

"What's that?" I ask.

"Demerol," a voice responds. "We need this room, so we're moving you. Just calm down."

Demerol? I think. *An opioid?* I'm so tired and weak that I let the nurse take Emma for the night. Cliff doesn't seem concerned either way, so off she goes to the nursery.

As the hour ticks later and later, family members trickle out, heading home for some shut-eye. Within an hour of the birth, everyone is gone, and all I'm alone.

As the drug invades my system, I realize I'm moving. My gurney is being wheeled down a hallway, and I'm floating, already high. I try to keep my eyes open and stay alert, but my lids droop. I'm on my back, unable to move on the thin gurney. A nurse wheels me into a utility closet between shelves lined with cleaning products. I know I'm not supposed to be in here, but I can only mumble gibberish. The nurse turns off the lights and leaves me there in the dark.

I can't move my body. I feel tears streaming down the sides of my head. I'm engulfed in fear. *What am I doing in here? Where is my family? Nobody knows I am in here!* I try to scream, but I cannot make a sound. I only know I am alive because I feel that my ears are full of my tears.

After a while, the nurse reappears. She flips on the overhead light and touches my stomach, seeming to make an urgent decision. Without acknowledging me at all, she pushes hard with both hands on my stomach, and I feel a warm gush of liquid between my legs. *I'm bleeding out,* I think. I try to talk to her, but she doesn't look at me. She must think I'm unconscious. She turns off the light and shuts the door again. I pass out in the dark.

I awaken before sunrise in a hospital room with two beds. A woman in the opposite bed is sleeping soundly. I blink my heavy lids. On the table next to me, I can see a pink plastic water jug, so I pour myself a cup of water. As I sip, I look for the drawer where my mom said the baby would be.

There is no drawer. Where is my baby? I thought the nurse would have brought her back from the nursery by now. I marvel at the strange new sensation—a primal sense of needing to know where another being is at all times, to have her near. This feeling that has just clicked on is a feeling that doesn't come to me from an idea, thought, or decision—it's maternal. Hormonal.

I assumed so much about childbirth, I realize. This place is chaotic. Not at all the three-day spa weekend my mother made it sound like after she brought my little brothers home when I was ten and twelve years old. Maybe my mother's experiences were idyllic, or maybe she didn't want to share gory details with me since I was just a kid then, but I listened to her every word, and her descriptions formed my expectations of birth.

I need to pee, so I hoist myself up and walk to the bathroom, stabilizing myself on the walls and tables to get there. I decide to rinse off in the shower. The water feels heavenly on my skin, still humid from sweat. I dry myself with a towel, put my hospital gown back on, and inch back to bed. I feel human again. I press the call button to summon a nurse.

When one arrives, she scolds me. "Mrs. Mulford, I told you, you can't get out of bed on your own! Did you take a shower?" She fusses around my bed and pushes on my shoulders to lay me back down.

"I don't know what you told me," I reply. "I was unconscious. Where is my baby?"

"The baby is in the nursery."

Her words and attitude strike a nerve within me.

"I want my baby. Can you bring her here now?" There's urgency in my voice.

"Let me check. You have to wait for a new shift, but I can check."

"I don't care about a new shift. I want my baby. Can you please bring her to me?"

"I'll have to get permission. Just lie down and rest. And if you have to go to the bathroom again, press the call button. You are not to be walking on your own yet."

There's a phone beside the bed. It's only 5:00 a.m., but I call Cliff. He's staying at my parents' house in San Jose, about thirty minutes away. My mother answers.

"Please come to the hospital," I plead. "I haven't seen the baby, and they won't bring her to me. Mom, please, I need help. Can you wake up Cliff and tell him to come here right away?"

Then—relief—the nurse returns with Emma, clean and swaddled. I hold her close and don't let go until Cliff arrives, which he does promptly. When he takes her in his arms, something shifts as we hold each other's gaze. For the first time, Cliff and I share something profoundly beautiful. Our eyes meet over this tiny life, and in that moment, we are connected—in unspoken beauty, awe, and the unbelievable truth that we made something extraordinary together.

As the sun comes up, I take off the hospital gown and pull on my own nightgown. Outside, the halls of the hospital buzz with activity. Normalcy returns. The joy of the baby overtakes everything in me. My eyes cannot see her enough. My arms cannot surround her more wholly. She's curled against me, small and warm, her breath soft and steady. I can't stop looking at her. She's perfect, so precious. Nothing in my life has ever felt so clear—this love, this responsibility. There's a knowing that I have to get it right. I will provide for and protect her. She deserves everything. And in holding her, I suddenly see that I once belonged to someone this way too. I was this small. This worthy. This full of possibility. I want to protect her with everything I have—and I feel, for the first time, that I need to offer that same steady kind of care to myself.

Friends and family arrive in a steady stream, bringing gifts and taking turns holding newborn Emma. I make sure to give her away only for a minute or two, just enough time for a photo. Then I ask for her back. I want to feed her often. She and I are both learning how to breastfeed. I'm not sure I'm doing it right, but I know we'll figure it out if I have enough time alone with her.

When there is a lull in the stream of visitors, I tell Cliff I want to go home.

My mother overhears. "You want to go home?" she asks, incredulous. "You have another night here at the hospital for free. You might as well just stay here and take advantage of it."

"No, I feel fine," I tell her. "I just want to go home. Really, I feel great."

I haven't told them yet what occurred during the night, nor have I even processed it myself—everything happened so fast—but it's starting to come back to me now. I know I don't want to spend another night here. I am physically fine, and so is the baby. I want us home by dinner.

"You'll stay at our place," my mother says. She offers to help Cliff take care of me and the baby.

I ring for the nurse again and tell her I'm ready to check out. She offers the same skepticism my mother did, but I am unflinching. Finally, she sees everyone in the room supports me and, without further argument, heads off to start my discharge paperwork.

As we are visiting and fussing over Emma, an aide dressed in a candy striper apron walks into the room. She approaches me with outstretched arms, matter of fact, and motions for me to hand her the baby. I only hold Emma closer.

"What do you need her for?" I ask.

"The doctor has to do a final wellness check in order for you to be released," the woman says, too abruptly for my liking.

I want to keep Emma close. I feel uneasy, my intuition flaring up. *Keep your eyes on Emma at all times*, it says, and I listen. Based on the events of the night before, I have very little confidence in this hospital or its staff.

"I will go with you," I say.

I'm not letting Emma go anywhere without me. It's later in the day, and the drugs have totally worn off. I am sober and awake now and so polite that maybe they don't realize I'm serious. This baby is not leaving my sight. I hand the baby to the nurse's aide in the candy striper apron and lift myself from the hospital bed to join them. The aide lays Emma in a cart and wheels her into the hall.

As we head down the busy hall, the nurse's aide tries to talk me out of accompanying Emma. "You don't need to come with us," she says. "This is going to take at least an hour, maybe two. Sixteen babies were born here last night, and the doctor is busy."

"That's perfectly all right. I'm coming."

"It's against hospital policy. You really must return to your room."

"No, that's all right. I'll wait as long as it takes."

I do not yield.

We arrive in a large room filled with baby carts, each holding its own precious cargo—newborns sleeping or restless, swaddled in pink blankets or blue, all donning little white caps. Emma is wheeled off into the crowd of other babies. A tall, heavyset doctor with black hair and a kind face looks up from his clipboard, surprised to see a mother in the room. When the aide explains curtly that I want to wait here while my baby is checked over, the doctor says, "Oh, no problem!" Then he addresses me directly. "I'll examine your baby first. Have a seat. This will only take a few minutes."

Relieved by his kindness, I sit in the chair he offers. Cliff, who followed us and now stands beside me, isn't as understanding.

"I don't know why you're making such a big production," he mutters to me under his breath, clear annoyance written on his face.

I don't care.

"I'm not leaving her alone, and that's all there is to it."

Through a glass wall, I can see the doctor examining Emma. After a few minutes, he emerges, the aide pushing Emma along in her cart behind him. The doctor cheerfully explains the results of the check-up, and I listen, keeping an eye on the aide, who has taken Emma out into the hall. When she gets beyond my field of vision, I turn away from the doctor and indicate to him my intention to follow her. He mirrors my gaze, his face showing alarm as he watches the aide turn out of sight.

"Where is she going?" the doctor asks me, as if I should know.

My internal alarms blare.

I rush out the door without responding and catch sight of the woman still pushing my baby's cart down the long hallway. Other nurses and doctors walk past without reaction. The woman isn't walking, though. She's running.

I break into a run, too, despite the episiotomy stitches pulling at me and Cliff yelling after me, annoyed, "What are you *doing?*"

I catch up to the woman as she stops in front of a utility elevator with a big red X on the door. "Hospital Personnel Only," it reads. The door opens.

I grab both sides of the cart and face this stranger.

"You can't be here," she says.

"Where are you taking my baby?"

"To the lab for tests. You can meet me down there, but you have to take the other elevator. This one is for personnel only." She points to the civilian elevator at the other end of the long hallway.

"No, absolutely not."

I keep hold of the baby cart and step onto the elevator.

Cliff catches up and joins us in the elevator as its door begins to close. "Why are you causing trouble? Don't you trust the hospital?" he mutters with disdain, throwing up his hands and shaking his head as if he thinks I'm behaving erratically.

On opposite sides of the cart, the aide and I face one another, little Emma quiet, angelic, between us.

"I am the mother!" I say firmly.

The aide stares back at me over the cart. To my surprise, she fights back. "And I am the nurse!" she says.

Cliff huffs from the corner of the elevator. It's clear whose side he's on, and it isn't mine.

The elevator doors open, and we emerge onto the first floor. Directly ahead of us sits the hospital's main entrance. And exit. It's immediately clear to me that this aide never intended to take Emma to any lab. Just beyond, revolving glass doors spill patients and visitors out onto a circular drive, where cars jockey for pick-up and drop-off positions at of one of the largest hospitals in the San Francisco Bay Area.

Still gripping the cart, I cry out.

"Help!"

My voice cracks, then strengthens.

"Help!" I yell again, louder.

All around me, people turn to look but keep on walking. My eyes dart from face to face, my anxiety growing.

Finally, I find my full voice and shout, a decibel above the crowd, "Help! Security!"

Then a blur of attention surrounds me. A blonde woman approaches and asks what she can do. She doesn't get too close, perhaps thinking I might be crazy standing there calling for help in my pajamas.

"Please," I answer, my voice quavering again. "Can you please get security? Something is not right here. I need help!"

Cliff stands by, visibly irritated with me for making a scene. Two mall-type security guards arrive, looking confused, and radio for backup.

Meanwhile, the aide has released the cart and slipped into a payphone booth. I see her pick up the receiver and dial. A moment later, the head maternity nurse and a suited man emerge from the elevator and head straight toward me.

"What are you doing down here?" the head nurse says, uncertain. "You and the baby are *not* allowed to leave the maternity floor!"

Too confused to speak, I allow her to guide us back onto the utility elevator. As the doors clamber shut, I look one last time to point out the woman in the candy striper apron but see only an empty phone booth.

She is gone.

Back in my room, I lay out the strange turn of events for the head nurse and a pair of gray-suited men with clipboards. My mother is there, too, with Aunt Caroline and Uncle Joe, who've stopped in for a visit. Visitors and investigators alike listen, incredulous, to every detail.

I recount the facts: How the candy striper took my baby. How she didn't want me to come with her. The doctor getting confused when the aide took off. Chasing the woman down the hall and forcing my way onto the utility elevator. The candy striper disappearing in the confusion at the main entrance.

Everyone in the room looks back at me, stunned, their jaws agape.

"We would never have let you get onto that public elevator," the head nurse tells me plainly. "Neither you nor the baby is allowed off this floor under any circumstances." My breathing turns shallow. The candy striper was violating hospital policy by taking Emma off the floor, and she was sending me down the hallway toward an elevator she knew other nurses would have blocked me from using.

Oh my god, I think, realizing the gravity of what has happened, *I've just thwarted the attempted kidnapping of my baby.*

"Please describe this person who took your baby," says one of the men with clipboards. Unsurprisingly, the "candy striper" isn't a hospital employee.

The two investigators take notes and make me feel heard. I want to go home and enjoy my beautiful child, but I also feel a desire for justice. Surely, I should report this to the police. The hospital investigators, however, convince me they will alert the authorities and follow up with me. I feel in that moment that action *will* be taken, but I realize later that, of course, the investigators are hospital employees. Their job is to avoid a real police investigation, which could result in a lawsuit and a lot of bad publicity. In my bubble, I just want to go home with my baby, and I am happy to leave this in their hands and walk away.

It's a nightmarish episode, but it teaches me a priceless lesson. I know now, with certainty I've never felt before, that my fierce love for my child and my instincts as her mother will supersede everything else in my life. Against any odds, I will choose her. I will fight to keep her safe.

1975—SUNNYVALE, CALIFORNIA

At dinner one night in our apartment at Grandma's fourplex, Sal looks up at the sad clown painting over the fireplace and says, "That's got to go."

Evan hides his peas inside his potato skin, and Sal smacks him on the side of the head, one more lesson in Sal's quest to teach us "discipline" and "respect." Just as casually, he joins my mom on the couch and watches TV. Evan stays at the table and stares at his plate for an hour.

I cannot believe it. My mom is snuggled up with Sal Caputo on the couch like nothing is wrong! She's so into him. I know then, once and for all, that we have to get rid of this guy. My mom cannot marry this one.

I talk with Lisa Reed about it the next day at school.

"You gotta be real bad," Lisa says. "If you're bad, he won't like your mom, and he will go away."

I nod, pretending Lisa's advice is sage, but there's no way I can follow it. Be bad in front of Sal Caputo? I'll only earn a concussion.

I decide instead to talk things out with my mom. Maybe she'll rethink her choice of partner if she knows how I feel about him.

"I don't like this man! Please get another one!" I beg her one night before bed.

I don't have the words to explain why I feel the way I do, but it doesn't matter. She doesn't have any interest in what I have to say.

"We're already engaged," she says, ending the conversation.

During the engagement, Sal hosts parties at his condo with some of his friends who make wine. Most of them are really nice. One of his oldest friends is Devorah, who lives in a neighboring condo. Dev is nice to us kids the way mistresses often are. I know about mistresses because that was what my mom said about Claire. Turns out Dev is Sal's ex-girlfriend, but now they are "just friends and neighbors." Right. I hear enough to know Dev is going to be out on her butt very soon. My mom insists on that, and for once, she isn't a complete pushover. Go, Mom! I like seeing my mom stand up for herself, even if I would prefer Sal Caputo going back to Dev and leaving us alone.

I overhear Sal say he wants to marry Mom because he knows she can have blonde-haired, blue-eyed kids. Though my mom's hair is dark and rich, I have blonde hair, pale blue eyes, and skin so fair you can practically see through me. Against my light eyes, my too-pink eyelids dominate my face. Kids at school call me albino. I looked the word up in the encyclopedia and found a photo of a white bunny with pink eyes. The only kind of bunny that isn't cute.

My mom's engagement means it's time to start going to church. Sal is a Catholic, and from him, Evan and I learn that we are living proof of

Mom's sin. If it weren't for us, Mom could have gotten her marriage to our dad annulled. Then she could take Catholic school classes and be married as a Catholic. Instead, we will have to stay in our seats when all the other congregants in church take communion. It's so embarrassing. I don't like church one bit!

"You are all sinners!" the father of the church yells. I look around at everyone in the audience. *Are you guys going to let him say that to you?* I wonder. But the crowd sits quietly, just staring up at him as if he said, "Look, the sun is shining!" I realize I'm the only outraged person here, so I fold my arms and slump down into the pew. How can this guy call everyone here sinners? He doesn't know anything about us!

During the divorce, my mom and dad split up the real estate they'd acquired together, and my mother kept the properties they had once intended as a nest egg for Evan and me—meant to help with college or someday purchase our own first homes. She confided in me, anxious about how she could afford the mortgage, taxes, and insurance. Worry consumed her. Since she's engaged to be married to Sal Caputo, she works up the courage to ask him for a loan so she can pay the taxes. He listens to her plea, appearing thoughtful, but really, he's just relishing the moment. I will come to know just how much Sal enjoys the rush of someone less fortunate seeking help from him, a powerful man with money. I'll see him do it to my aunt Brenda and uncle Fred when they ask for a loan to save Uncle Fred's mechanic shop. He'll quietly strip them of their dignity as he makes them grovel and wait for his decision, which is inevitably a "no."

With that same routine, he tells my mother he'll need some time to think over her request. For days, my mother wrings her hands, waiting for his answer, kicking herself all the time. How did she let her finances get so bad? With a little help, she can surely make things work. She has to.

You can guess what happens next. Sal doesn't just reject her request. He shames her for it, reminding her who holds the power in their relationship. "It wasn't right for you to ask this of me," he says, "but I forgive you." He *forgives* her—but he won't marry her knowing she carries debt. Just like that, she sells both the houses and declares bankruptcy, and Evan and I say goodbye to the houses that were supposed to pay for our college educations.

Eventually, my mother tells me she had no choice but to sell and declare bankruptcy—something she said Sal insisted on before he would agree to marry her.

Evan and I spend the week before Mom and Sal's wedding with our Dad. It's summer, and we're excited for a long visit with him. One morning over silver dollar pancakes, Dad comments on my long hair.

"Don't you think it would be nice to get a shorter cut for the summer? It'll be much cooler," he says, forking a mouthful of syrupy goodness.

"I would love that," I reply. "Sounds great!"

"Let's do it today!"

After breakfast and some chores around the house, Dad fishes a pair of scissors out of the bathroom, and in three or four cuts, he chops off my long blonde locks. In the end, I sport a fashionable bob, which I love. It feels so much lighter, fresher, and cooler. Plus, it makes me feel a little bit older. It's a very grown-up change for me, perfect for moving into our new house and getting ready for fourth grade and all the friends I am sure to make on our new street.

Mom doesn't feel the same. When Daddy drops us off after our visit, she totally freaks out, bursting into tears at the sight of me. This makes Sal blow up too. He screams obscenities while she cries.

Meanwhile, I retreat into shame. Mom's reaction makes me feel ugly. Ugly and wrong. She's acting like the best part of me is gone. Like I've ruined her big day. It's a disaster I hadn't anticipated.

My mom and Sal Caputo Jr. get married in 1976, one month before Daddy and Claire's new baby, Bryce, is born. Exactly nine months after her wedding, Mom gives birth to little Giovanni Salvatore Caputo. Now, it seems like my mom wanted to get married even faster because Daddy got married and had a baby coming already. Mommy needed to catch up to them.

Photos from the wedding day show my stylish bob hidden underneath a wide-brimmed yellow hat, a scowl etched across my face. Right up until the vows, I silently hoped they wouldn't go through with it.

When my mother shares the wedding album at dinner parties, viewers inevitably ask my mother why I look so unhappy.

"Oh, she didn't want to wear a hat," Mommy says.

She won't admit it's because I don't like her husband.

CHAPTER FIVE

SIGNED, SEALED, RECONSIDERED

"Everything will be okay in the end. If it's not okay, it's not the end."

John Lennon

1991—SUNNYVALE, CALIFORNIA

Once we're finally home with Emma, my mother and Sal start scheming. Midway through the pregnancy, Cliff and I moved out of our one-bedroom loft and rented a house in Capitola from Dorothy. We adopted a dog from the pound and bought furniture with the money from the wedding guests. But now with the baby here, they want us to move back to San Jose so that we'll be closer to them. My mom is in a not-so-subtle competition with Cliff's mom to decide who will claim the coveted title of grandma. Mom has her way. Cliff's mom will be "Grandma Dorothy" while she'll just be "Grandma," the original.

Sal gives Cliff an accounting job at his land development company, the most prolific developer of multifamily apartments and townhome communities in Northern California. Now, my parents are offering us a brand-new three-story home in one of their townhouse developments so I can go back to work part-time and my mom can help me with the baby more easily.

In return for the lifestyle upgrade, Cliff and I will watch over the new townhouse construction after hours. Sal claims that he'll eventually put the beautiful townhouse in our names if we prove responsible over time, but Sal himself picks out the color scheme and carpet. Everything must be white. I get a job working four hours a day, three days a week in a small optometrist's office in a strip mall in a sleepy area of Sunnyvale, and I find a lucky spot with a nanny to care for Emma. The nanny, Hermione, has raised a houseful of babies for a neighbor whose children are older and now in school. Hermione has the time and loves having a little baby to dote on. Emma's in excellent hands while I'm at work. Cliff and I can afford to pay a moderate rent to my parents, who are now both our landlords and my husband's employer.

They're being supportive, I tell myself. *This is what parents are supposed to do for their kids.* As soon as Cliff and I agree to the arrangement and sever our ties to Dorothy's house, my parents start insisting on a few new rules, which they reveal seemingly at random. First, they decide they don't want a dog in the house. Our puppy has to be rehomed. Goodbye, my sweet yellow lab, Bones. Then, Mom breaks the news that our cat isn't welcome either. I try to fight this one. I've had my cat, Kia, since I moved out of their house at seventeen. I've been through a lot with that kitty, and I love her. Eventually, my mom's brother, Uncle Mark, his wife, Aunt Lauren, and their young boys take her in, renaming her "Bullet."

Shortly after a very showy housewarming party at which Sal ironically spilled a full glass of red wine on the brand-new white carpet he had insisted on, we finally begin to settle in. I find my strength coming back. On the days I'm not working at the optometrist's office, I take baby Emma to the park or for swims with other new moms I have met.

I do my best to manage under the controlling behavior I have experienced with Cliff since our wedding night, but I start dreaming about returning to the classroom. I mention this to Cliff and beg him to take weekend day trips up to San Francisco and Marin County to experience some exciting new areas. I planned weekend drives up to the North Bay, Sausalito, and a town I fell in love with: Petaluma. I began to dream more realistically of chickens and land, a farmhouse to raise a family. My budding dreams that ignited when Emma was born had no appeal to Cliff, but I couldn't stop thinking about them. I dreamed of having a garden and chickens. A farm. I tell Cliff I've looked up the salaries for court reporters and want to finish my training that I began during my pregnancy.

"I could make the same as you do in my first year!" I tell him, excitedly hopeful. "And the salary goes way up the more seniority you get. Can you imagine if we both had great incomes? We could build our savings so much faster!"

"Ha!" Cliff laughs at me in the kitchen after work one evening, "You are not *smart enough* to go to school." He drops the idea. "It's off the table. I'm the boss around here, and I won't have it." He's not leaving any room for me to follow up. I know he knows I *am* smart. Does he really mean he does not want me to surpass him?

One day, my parents ask me to come to their house for dinner—with Emma but not Cliff.

"What's been going on with Cliff?" my mother asks across the dinner table.

I'm not sure what to say.

"Nothing," I finally answer. "Everything's normal."

"Has he been sick?" Sal asks, chewing his food.

"No, not at all," I answer. By now I am growing suspicious.

"So he's been going to work every day?"

"Yes, same as usual. Why?"

The two exchange a knowing glance.

"Cliff hasn't been to work in over two weeks," Sal says. He gazes at me with piercing eyes, and I feel like *I* am the one who hasn't shown up for work.

I listen, dumbfounded. Cliff has been getting up in the morning, putting on his slacks and shirt, and coming home after 5:00 p.m.

I'm stunned. This level of dishonesty is incomprehensible. What was he doing all those days? What kind of person can lie like that? What kind of person believes he can get away with it? What kind of person am I married to?

Back at home, I ask Cliff what's going on. He breaks down crying.

"I hate my job," he blubbers.

He explains that everyone in the office is expected to take turns answering the phone at the front desk for an hour once a week while the receptionist is on her lunch break. Cliff refuses.

"It's beneath me," he says through his tears. He thinks answering phones is a woman's job, so he's humiliated by it.

I'm taken aback, but I know questioning this point won't do any good. His view of a woman's role is not my most pressing problem at the moment. Instead, I reframe the problem. Cliff is stressed at work. He'll be a happier person and a better partner and father if he doesn't hate his job.

He doesn't want to be an accountant anymore, he says. He dreams of becoming a baker.

A baker? Oh, wow. In the short time we've known each other, he's never shown any interest in the kitchen, so I'm surprised. I hesitate, but okay, well, what can I say?

I push down the alarm I initially felt upon learning Cliff had been lying to us all for weeks, and my focus shifts to an overwhelming question forming in my mind—how would I survive a divorce? I can't yet entertain my original desire to leave Cliff before I got pregnant and married him. I can't admit to myself that I was so mistaken, and I feel like I'll be discovered as a fraud for pretending to love him and accepting all the presents, the house, and the attention of my parents. I worry everything will disappear if I don't support him. So I bury it all and try to see something good. A turning point? Well, it would be a big change, but any change seems better than things staying as they are.

We agree that I will go full-time at my optical job in Sunnyvale, and he will apply for a bakery job at the Whole Foods on University Avenue in Palo Alto. His shift will run from 3:00 a.m. until 9:00 a.m., and then I'll go to work at 10:00 a.m. Good. I will be relieved of the morning sex shift. He'll be Emma's primary caregiver during the day, and we'll let Hermione know we no longer need day care.

At about 9:15 each morning, Cliff returns from work, raisins stuck in the soles of his shoes, tracking flour and icing all through the house and grinding raisins into the white carpet. Our sofa is soon covered with a thin coat of flour when Cliff plops down every morning to watch daytime TV, cracking open a beer since it's the end of his workday, after all. I'm concerned about Emma, but he is her father. Perhaps the change will better connect him to that role. Plus, being married to a baker has its perks. Whole Foods is a wonderland of luxury foods I am just beginning to discover.

One day, we forget to move the baby's car seat from my car to his during the morning handoff. When I get home from work, I notice his car is parked in a different parking spot than it was when I left that morning.

"Did you go somewhere today?" I ask, Emma on my hip as I start to prep dinner.

Cliff shrugs and explains how he ran out to buy vodka and cigarettes, leaving six-month-old Emma home alone.

I am thrown by this news.

"You left her alone?" I put Emma down in her playpen.

"It was just for half an hour. She was in the crib. What's the big deal?" He sounds like a teenager defending himself to his mom, not a grown man taking responsibility.

"What if there were a fire? What if something happened, like you got a flat tire, while you were out? You can't leave a baby at home alone!"

"You are so overprotective! Don't tell me how to take care of my daughter. I'll do what I want with her!"

And that's it. The last straw.

At the beginning with Cliff, I somehow mistook control for commitment. It was easy to believe that someone who watched and criticized my every move must care deeply. But over time, I saw the difference between care and possession. I was not safe in that dynamic. I was studied, assessed, judged, and shamed.

At once, I mentally add up everything I've been tolerating—the forced sex, the beratement, the constant accusation that I'm frigid, the lies when Cliff ditched work for two weeks to drink vodka in the park. It's enough. Add to that Cliff's careless endangerment of our child, leaving her at home alone? It's the moment I finally decide to exit this marriage. Immediately.

I'd been conditioned by the treatment I received as a child to believe that I was powerless. My mother had instilled in me that whenever I was being treated poorly by her husband, it was my fault, not his or even her fault. I did not act earlier because I felt that if I left my marriage, I would be outcast more than ever before, labeled a troublemaker, and deprived of the emotional support and acceptance that I'd recently enjoyed from my parents. I anticipated having to return the money and gifts we received at the wedding, and who knows what else.

I had little choice but to depend on my parents and Cliff for so long, but now, working outside the home allowed me to begin to finally establish my own autonomy and imagine a peaceful, self-sufficient life. I felt emboldened and capable now that I had my own paycheck and my body had regained its strength and returned to its normal state post-pregnancy. I was subconsciously preparing for a new era, and with this latest instance of Cliff's alarming and escalating destructive behavior, I know I have to get out this time and that Emma and I can survive without him.

Cliff, thinking himself a king, believes that I should be the one to leave the townhouse. The townhouse my parents own.

"They like me more than they like you," he taunts.

Yes, my relationship with my parents has always been rocky, and, yes, they have warmed to me since the birth of their first grandchild. Cliff uses these facts like a weapon, implying that without him, my parents would go back to their disapproval of me, and his words touch some vulnerabilities that exist in me from childhood, just like he intends. But I know, without a doubt, that even my perpetually unreliable parents won't go that far. They won't let him go on living in the townhouse without me and the baby. That's just not logical. Still, he refuses to leave.

Cliff has made a couple of friends at the bakery and has been meeting up with them to watch sports and drink beer on the weekends while I stay home with Emma. The more alone time I have, the better I feel. Cliff and I agree that him staying home with Emma all day during the week after his early shifts is not working. Now, I *have* to find a way to get Cliff out of the house for good. Except for arguing about finances or household chores, we ignore each other at home, but I am growing more and more uncomfortable with having him come and go.

Finally, my mother and Sal, with their two young sons in tow, come to kick Cliff out of the house. As soon as they're through the door, Cliff looks at them, reads the writing on the wall, and, without any words, grabs his bag and walks out. When tested, he ultimately fails.

Cliff gets his own apartment, and my parents call in a ruthless legal ace, Link Maddox, who handles our uncontested divorce quickly and effortlessly. I offer Cliff all our material possessions and don't ask for any alimony or child support. Custody of Emma is the only thing I want. Cliff agrees without protest. After all, he's already dating and hardly wants a baby around.

In Link's office a month or two later, I sign the last papers, and Sal escorts me out of the attorney's office. As we descend in the elevator, I feel both relieved and protected by my parents, something I've always craved. Our relationship is more solid than it's ever been, and I have a new life to look forward to.

But I've forgotten how easily Sal can slaughter hope.

"You don't even realize what you have done, do you?" he says, shaking his head.

"What *I* have done?" I parrot.

Our eyes meet across the elevator.

"You've really fucked that little girl of yours. No one will ever love a child who isn't theirs."

I know the truth in his words because I've lived them. It's no secret that the life I've shared with my mother and Sal has been rough. As the elevator dings past floors, I recall how Sal pushed my real father away, along with all the other caring adults in my life.

1978—SAN JOSE, CALIFORNIA

One hot summer day, our grandfather, Baba, picks up Evan and me instead of Daddy. The best thing about Baba is that he always has a tin of Strawberry Quik and a fresh quart of milk for us when we stop at his house. A dapper, pipe-smoking, old money gentleman.

Baba is taking us to meet up with our dad at Great America, a theme park featuring cartoon characters we know from TV. We spend the morning riding merry-go-rounds. When Daddy arrives, he sits me on the merry-go-round's prettiest white pony. As we spin 'round and 'round, Evan and I wave, but Daddy and Baba are deep in conversation near a bench.

When we tire of the park, we head off to Farrell's for banana splits. Dad sits in the leather-tufted booth, smartly crossing one long leg over the other as we dig into our ice cream treats. Then he presents each of us children with a box. My box is velvet and holds a delicate golden quartz watch. I squeal with delight. Evan gets a toy truck, and Baba takes him to play with it outside.

Daddy secures the clasp of the new watch on my wrist and looks into my eyes. "I'm giving you this present," he says, "and I'm asking you to give me a favor in return."

He has a secret that I must keep. The watch is a reminder of that secret.

I promise.

"Claire, Bryce, and I are moving to England, where Claire is from. She wants to be closer to family," he tells me. "In fact, Claire and Bryce are there already."

A cool drop of ice cream slides down my chin.

"I'm leaving today," he continues.

I'm in shock. I have questions, but they're too big for my child's mind to grasp, let alone verbalize.

Finally, I ask the best question I can think of.

"Where will Mochi go?"

I'm desperate to know the fate of that lovable poodle, my companion since the day I was born.

"We gave her away," my father answers.

I know how to be devastated over the loss of my dog, but I can't grasp the concept of losing my father.

Evan and I ride in the backseat of Baba's Cadillac, my dad staring straight ahead in the passenger seat, as we head toward San Francisco International Airport. The gold watch glints on my wrist. At the airport, we walk with Dad to the gate. I don't remember hugging him goodbye, but I do remember sitting by the big spectator window and waiting for the plane to take off. I tried to make out my dad through the small, round plane windows.

"I see him," Evan says. "He's looking at us and waving."

My dad's words were ringing in my ears.

"We won't see each other again, Susie."

There are no plans to visit, only a gold watch to remind me to keep this secret from my mother—and everyone else. I can't tell anyone that my dad left me and that my brother Bryce is already overseas without me even getting a chance to say goodbye. My dog Mochi belongs to a stranger now. I don't understand.

Baba drives us home. Our garage door is open, so I walk through the garage and into our living room, sit on the couch, and watch Baba drive away. Finally, I burst into uncontrollable tears.

My mother, pregnant again with her second child with Sal, sits down next to me. "What's wrong?" she asks.

I look at the gold watch, then at her. Through heaving sobs, I choke, "I promised Daddy that I wouldn't tell."

My mother puts her arm around me and lets me cry into her blouse. Then she looks me in the eye and calmly says, "You can tell me. I'm your mother. You're not supposed to keep secrets from me. I love you. Everything's going to be okay. Just tell me what's wrong, and I will help you."

My mother is being so sweet, and I really, really need somebody to comfort me in this moment. I don't understand what's going on. I don't understand what it means to never see my dad again. I don't understand how far England is. I don't understand why he said they're not coming back. I don't understand how they could just give away my dog. I'm terrified, so I tell my mother, not ten minutes after my grandfather pulls out of the driveway.

Betraying my dad reignites my uncontrollable sobs as I look at my mother, desperately needing sympathy. My expectation that she will be equally outraged about Mochi's fate goes unfulfilled. Instead of soothing my pain and alleviating my guilt, my mother forgets about me and jumps up. She runs to the phone in the kitchen and dials while yelling for Sal through the kitchen window. He's in the backyard working on his wine.

I've stopped crying. I'm frozen in place on the couch, seemingly invisible as my mom immediately shares my secret with Sal, who curses and calls my dad a bastard. He rips the phone out of her hand to speak to the person on the other end, a lawyer. While I sit there, tear streaks drying salty on my face, they discuss how to put my dad in jail.

I get up and go to my room without anyone noticing. I lie on my bed and cry myself to sleep. When I wake up, I take off the beautiful gold watch and trace the deep creases left on my arm with my finger. I have cried all the tears in my body, so now I have no choice but to do normal stuff like go to the bathroom, brush my teeth, do my chores, and go back to bed like any other day.

After a couple of weeks, I overhear my mom and Sal talking about how it is actually a good thing that my dad is gone. Even though it means no more financial support, at least they won't have to deal with him anymore. My mother is angry that my dad left for England, not because of the pain it caused Evan and me but because he is getting out of paying child support. My mother says one day, "How unwanted you must feel . . . Your father disappears with your half-brother but abandons *you.*" I am taken aback by this brutal statement said out loud. Why would she say that to a child mourning a parent's loss?

Now, to Sal, our abuse is even more justified. Instead of holding back, he has a reason to be disgusted with Evan and me. Sal Caputo seems to enjoy the fact that his resentment of us now has a rationale and that there will be no adult who will listen to us. Every nurturing grown-up has been eliminated from our lives.

"You should be grateful that a man who is not your father is taking care of you," my mother reminds Evan and me. I'm reminded of the family dynamic. She's Sal's wife, and my younger brothers Giovanni and the new baby about to arrive are his biological kids. He is their real father. Evan and I are just lucky that we have food on the table and a roof over our heads.

"We are your only parents now," Mom says. "You must never ever tell Sal that he is not your real father because that will hurt his feelings."

"But he always tells us he is not our real father," I protest.

"You must be respectful," Mom says, shushing me.

In my mind, since Evan and I were already alive when Sal married my mom, we were part of the deal, right? He could have married someone else if he didn't want us, right?

It's not my fault that my dad left, I want to tell her. *It's not my fault that you got divorced.*

You told me it wasn't my fault. Remember that?

CHAPTER SIX

REBOOT

"It takes courage to grow up and become who you really are."

E.E. Cummings

1992—SAN FRANCISCO, CALIFORNIA

After divorcing Cliff, my husband of just over one year, I move to San Francisco, a single mom to one-year-old Emma. Having already reached the pinnacle of the opticianry pay scale at fifteen dollars per hour, I need to find a more lucrative career so I can support us both, and I'd already been dreaming about continuing my court reporting training.

The plan for my relocation comes with the help of my mother and Sal after Cliff continues to show up unannounced at the townhouse at night, months after he's moved out, like our home is still his. It's a clear violation of boundaries, as if he still believes he has the right to barge into our home and exert his control over Emma and me. Cliff's behavior reflects a sense of growing entitlement and coercive control, and I don't feel safe.

One day, I muster up the courage to show my mother that the stress has been coming out of me in the form of eczema and scratching at my skin. I don't understand it, but I know I have to tell someone and get help. Thankfully, my mother takes one look at the scabs on my arms and tells Sal that Emma and I need to get out of there.

I pitch an idea to my mother and Sal: I get back into court reporting. I find a fully accredited trade school in San Francisco where I can test into the two-year program since I'd already completed the initial training back in San Jose and hit the ground running, complete my training, and obtain the certificates needed to have a well-paid, secure career in the courtroom in about one year's time. They put the townhouse on the market, and I start looking for apartments in San Francisco.

Something amazing happens when I'm in the city. I start to feel like I belong somewhere, right here—San Francisco, the city by the bay. It's where my mother was born and where, over fifty years before I take up residence, my teenaged grandmother Emeline walked the Golden Gate Bridge on its opening day.

Weird, creative, and full of contradictions—just like me—San Francisco soon becomes my inspiration for the unconventional. It's the perfect home for misfits, a place bursting with character. Sensual, strikingly beautiful, emotionally raw, deeply original, and unapologetic. The city becomes my friend and protector.

At first, I am cautious navigating the hills with Emma on foot, both of us sensitive to the defiantly unpolished energy on the streets. But when I take a moment to tune out the noise of the cars and the buses, I find my reward in looking up at the gargoyles in the eaves—those carved stone muses guarding the skyline. The city is full of incredibly fascinating architectural details that go unseen unless one stops to look up.

Emma and I move into a one-bedroom apartment in St. Francis Place facing a gray office building and a deserted industrial alley in the South of Market neighborhood. Our two windows see no sun, but I love that the complex is filled with families far from home and children growing up abroad. In the center of the apartment buildings is a concrete park where children can ride bikes and roller-skate. There are playgrounds, an outdoor swimming pool, a concierge, a gym, and a lovable doorman named Curtis who stands at the top of the pyramid of steps above Fillmore and 3rd Street to monitor everyone coming and going—Cliff will not be able to show up here without permission. The community room off the pool serves an unlimited, free breakfast buffet every Sunday morning with delicious French pastries, coffee cake, blueberry muffins, scrambled eggs, bacon *and* sausages, fresh fruit, coffee, and juice. Children play as parents sip coffee and visit with one another. I am the only one in the group born and raised in California and the only single mom. Most of the other residents I meet are European couples who relocated to San Francisco for their careers.

I know how to make Emma's favorites—spaghetti Bolognese, grilled cheese, beef and cheddar loaded tacos with real tortillas and pico de gallo—but often our dinner is Top Ramen with toast and a soft-boiled egg. It's deliciously comforting but also cheap and always in the cupboard for those weeks when I can't quite afford a trip to the grocery store.

Emma started walking at eight months old and, by the time she is a toddler, already has an incredible vocabulary. She's able to hold highly complex conversations with adults, but her baldness makes her appear younger than she is. People are constantly surprised by her advanced communication skills. She's an adorable baby, walking and talking, dressed in pink jeans and skater T-shirts or floral dresses with black Converse high-tops. I spend too much on colorful Doc Martens, flower-adorned hats, and miniature sunglasses. Every parent may think their child is wonderful, but Emma is *magnetic,* a baby with an old soul who truly cares for friends and neighbors. I marvel at her constantly, knowing she has much more to teach me than I can ever teach her.

Every morning, Emma holds my hand as we walk to school. She becomes fast friends with Conte, the homeless man who sits with his cup on the corner across from her school. Since I'm busy watching traffic, I am initially unaware of their shared waves and smiles until Emma mentions her "new friend Conte on the corner." After that, I start waving to Conte too.

Emma receives a free donut from the coffee cart vendor on the corner opposite Conte. In the afternoons, she graciously receives a piece of penny candy from the tall and handsome Pakistani brothers who work at their father's bodega below the stairs of our apartment building. I love San Francisco, but this is clearly Emma's city.

Trinity Business College, the court reporting school I attend, is conveniently located right next door to Emma's day care. Because I've already had some training, I'm able to skip the first year of introductory theory and go straight into speed-building, accuracy drills, and the academic courses required for certification—advanced English, medical and legal terminology, and transcript preparation for courtroom work.

At school, I meet other young women my age: Marla, Yvette, Misty, and Lorena. All of them adore Emma and come up with excuses to spend time with us both. I also reconnect often with Ramona, Ann's best friend since high school, who also recently moved to the city and works as an editor at *Parenting* magazine. I introduce her to my Italian neighbors, Vincent and Federica, who have two wonderful children Emma's age, and we all become close.

These women become Emma's aunties and help me raise her as she grows. They babysit so I can go grocery shopping without wrangling a stroller, a child, and grocery bags on and off the bus. When I ask Emma who she'd like to invite to her birthday party, she surprises me by dictating a list absent of any of the kids in her preschool. Instead, she rattles off the names of my friends. Seeing my puzzled face, she explains, "They're my *adult* friends." But of course!

At Trinity, I become infamous for having a "computer machine." Back at the vocational school in Almaden Valley, our teacher, Hilda Fenton, taught me an elevated court reporting shorthand language on a futuristic electronic court reporting machine that had a motherboard, an LCD screen, and a thermal heat transfer printer instead of the ecru-colored kraft paper tape and stamped ink keys of the traditional machines everyone else was still using. The computer inside my machine, paired with the elevated shorthand language, produces English words in real time on the thermal paper output, and the machine can be tethered to a desktop computer to transfer the content instantly without the laborious step of typing the translation by hand from shorthand code to English.

This instantaneous English output is a game changer in the court reporting field. In the beginning, though, I'm seen as cheating because I skip the majority of the work involved in taking traditional shorthand tape and manually translating it to common English at the typewriter banks after each dictation practice session. My machine and the language I've learned eliminate the need to translate and type. Instead of a typewriter, I use a desktop computer.

After some time, I am able to convince the school board that I'm not cheating at all, but rather that my process is the way of the future. I have done the work up front by learning a much more difficult shorthand language, enabling me to produce a clean, legible output with minimal editing.

However, I hit a plateau in my speed, getting stuck at 180 words per minute with 98 percent accuracy. To graduate and take the court reporter qualification tests, I have to consistently achieve between 200 and 220 words per minute. Many students stay in the advanced speed rooms, practicing like endurance racers for months until they reach the required speed and accuracy benchmarks, but I just can't force my speed to progress. I need more time and practice to get to the professional level—time I don't have. My parents agreed to support me for one year, a swiftly approaching deadline. Naively, I thought they'd see my progress and give me more time, but no. Now, they're about to rescind their assistance with my rent and expenses. I have a Pell grant and a student loan, but money is running out, and I don't have enough of it to sustain Emma and me. I haven't received any child support since the divorce and have no other means of income.

One day, close to my one-year expiration date, I walk into a presentation at school. A local video post-production engineer is speaking to a dozen students about a transcription job. He's looking for people who can type fast on a computer program called closed captioning for television. I am immediately intrigued, take his business card, and make an appointment to meet his boss at a studio called Fast Forward Duplication and Post Production.

In a boxy black-and-white herringbone dress with black heels and pantyhose, I walk up Montgomery Street from St. Francis Place through the Financial District, pass the Transamerica Pyramid, and find the Fast Forward studio while wheeling my computer machine behind me.

Stepping into the dark post-production studio feels like entering a video game den of brothers. I come in confidently and shake the boss, Bill Kinder's, hand. There are a few of us there for the training session, and he demonstrates how the program encodes closed captions into the video signal for prerecorded films and TV commercials. I ask a lot of questions to understand the work, while the other potential transcribers behave as if their parents made them show up. No one is excited but me, and I am in top form, sizing up the competition and planning how to strategically pick them off so I'll be first in line. This flexible on-call work could be my lifeline, allowing me to stay in school and keep my apartment if I get enough reliable hours.

After the demonstration, Bill announces that we can come in for practice to determine our ability and interest. Then we'll share the work —between the four of us. I don't want to share any of the work. I stall until the other job applicants have left, look Bill in the eye, and ask him for priority.

"I'll be the best at this work, and I'd like to take all the projects myself. Can I have your word that you'll call me first when new assignments come up?"

The look of surprise on his face doesn't faze me because there's a smile in his eyes. He stares back at me, unable to say no, and I detect admiration for my ambition. And, of course, not just my ambition, but also the desperate knowledge that I am now the sole provider for all of Emma's needs.

I do not disappoint Bill Kinder. I take every job he throws at me, from overnight deadlines on commercial ads and corporate videos to feature-length TV movies. If I quit court reporting school, I'll have to start paying student loans, so instead, I take a six-month leave of absence and arrange for Marla or Federica to sleep on my sofa and look after Emma at night while I work at the studio. I save longer projects like made-for-TV feature films for every other weekend while Emma is with Cliff.

I find Cheetah Systems, the developer of the captioning computer program, in the phonebook and reach out to them to learn every aspect of the tool so that I can be the definitive expert at Fast Forward, solidifying my job security. There are numerous bugs and glitches in the software, and I need to find workarounds for each of them to do the job well.

This works for a couple of months, but it's soon clear that between paying my own rent and Emma's day care, I can no longer afford the tuition for court reporting school. I need more work than Fast Forward can give me. I have to figure out a different way to support us. Desperation is my motivator.

Since the Cheetah Systems software was developed in Fremont, California, I take a BART train to meet CEO and CTO Gary Robson, a big, tall door-to-door salesman type of guy in a three-piece suit from Men's Wearhouse and with a bushy seventies mustache. Gary towers over me in my herringbone dress. There's no training program in place at Cheetah, but I convince Gary to sit with me for four hours and show me the inner workings of TurboCAT, the Computer-Aided Transcription closed captioning software. I'm determined to leverage my opportunity to make enough money to support myself and my daughter while staying in our apartment. I hope the computer program can somehow help me increase my court reporting speed outside of school so I can qualify for the certification test.

I've added quite a bit to my knowledge about this kind of technical software and encoding by the time I learn about the Americans with Disabilities Act, which passed in 1990 and requires 90 percent of all broadcast television to be closed captioned by the year 1994. I see the implications of the new law. Who else in San Francisco is set up to provide 24-7, full-service closed caption encoding to accommodate these new requirements?

I decide to start my own business. I am now twenty-five years old, still in the same one-bedroom apartment, still a single mom, and I've just heard the word *entrepreneur* for the first time. Initially, I discount this term as pretentious and excessive. I just want to work and make my own hours, but I still remember my dream for a piece of land in Petaluma. A farmhouse with chickens and a garden. How will I ever make enough money to buy land in Petaluma! Even a very small house with a tiny yard in Petaluma requires a giant down payment and monthly mortgage payments in the thousands. Even if I could get a home loan, I would have to secure a well-paying job and pay thousands each month. The only way I know how to earn a million dollars while I'm still young is to start a business. I imagine I can get enough work if I buy a computer like the one at Fast Forward and a $15,000 license for my own copy of the Cheetah Systems software. Then, I can work at home while Emma sleeps and get contracts during the day while she's at preschool, allowing me to finally think about growing my business and buying a starter home on my own.

I apologize in advance to Bill Kinder, thank him for training me, and let him know I plan to become his competition. With an inconsistent and waning number of captioning requests coming in, Bill tells me he figured I'd be moving on. "And I appreciate the heads-up," he laughs.

I ask Cliff for $250 per month in child support, which he refuses, so I ask the courts for help. My request becomes a court order fairly quickly, so I turn my attention to figuring out how I can possibly pay the insane sum of $15,000 for the software license. That's more money than I can even imagine in 1993.

I begin taking night school classes at a business incubator for aspiring women entrepreneurs—to learn how to grow my business, yes, but also so I can apply for small business grants and low-interest loans once I complete the prerequisite series of microenterprise courses. A teacher who admires my technical aspirations and is helping me hone my business plan pulls me aside after class one evening.

"I was recently contacted by Bank of Canton, a local bank in the Financial District catering to the Chinese community," she tells me. "They've been cited by the SEC for having too little diversity in their portfolio, so they're actively looking for white entrepreneurs to lend to in order to get the SEC off their backs. You should give them a try."

I stroll into the branch office on Montgomery Street with my business plan and get a $15,000 loan approved for the software in one afternoon.

I surely wouldn't have broken ground on this new business if I'd known how hard it would be just to get my first client. I don't even know how to budget a new industry segment. I just take it one step at a time, determined to succeed. Instead of researching the viability of my business idea, I just go for it, knowing this is my one shot. I smash every curveball and keep going.

After completing the obligatory classes required to qualify for securing an SBA loan for the additional equipment I need, it takes me another eight months just to get my first client. I land one thirty-second commercial, for which I'm paid $300. I can finally use the technology infrastructure I've leveraged all my resources and loans to buy.

Once I land that first client, others start coming through. I am suddenly working consistently, and money starts rolling in. Believing my business needs a boost to continue this momentum, I enlist my brother Evan, who is a prodigy in graphic design, to create a logo and flyer that I print off at Kinko's and hand-deliver to all the local TV studios, including KNTV, KQED, and KRON, as well as advertising agencies like Goodby Silverstein and Ogilvy and all the video post-production houses. I work my way into corporations such as Nike, Safeway, PG&E, and Sun Microsystems. Finally, my little company, CTV Captioning, is known and respected in the industry.

There are more barriers ahead, though. I soon realize that the computer and software I bought with the loan aren't enough on their own. Creating a time-coded video signal must be done in various formats, requiring expensive professional video decks found only in the equipment racks of post-production houses. I buy one video encoder that works with the most popular formats, but then I still have to try to convince a post-production house to install my device in their racks and create the duplicates for me and for my clients—their competitors! A wildly outrageous request.

I persist all the same and soon enlist a post-production company called Transmedia to house, maintain, and operate my encoder on demand for a small fee. I have to rely on them to put aside all their other work to help me meet my deadlines whenever I show up so that I can then distribute the master tapes back to my clients. I'm acquainted with the deer-in-the-headlights looks from video professionals by now, and I know Transmedia has no good reason to do anything for me. But they do it. They do! It probably helps that I won't take no for an answer, pestering them, calling them, and dropping in to visit until they agree. But "no" is always a possibility. Nevertheless, I am relentless, professional, and polite, and I believe they admire my impractical idealism and, on a human level, want to see me succeed. I negotiated hard, pushed my way into that role, and was proud of it.

I join forces with the Mayor's Disability Council in San Francisco, which, among other things, advocates for individuals with hearing impairment. They help me push the idea to the local media industry that closed captioning is not just the law but the future—and a benefit for them rather than just a requirement. Of course, they were way ahead of me, but it felt like they appreciated having a business owner show up at their monthly meetings. It was a surprise to them. I recall asking Gary Robson to do a presentation on Cheetah System technology for the council to help them with their mission even more.

One of my most consistent clients is the local PBS station, KQED. At night at my desk in my bedroom, I lay the captions for PBS documentaries and cooking shows filmed in San Francisco, enjoying this deep connection I am building with the city. It was an honor. I'd eventually work on Jacques Pépin's *Cooking with Claudine* and *Neighborhoods: The Hidden Cities of San Francisco*.

The production manager at KQED, Eric Dauster, cares about accessibility and fully supports getting all KQED videos and films captioned. In understanding the business side of how public television works, I learn the difference between the companies that want to serve the public and produce excellent work and the ones that only care about making money. At Eric's urging, I add live captioning to my company's services and hire high-speed, real-time deposition reporters to spend their Friday evenings working for me. We begin live news captioning that requires me to purchase another layer of software and additional encoders.

As I establish ad agency clients for national thirty-second commercials and big companies hire me to caption their monthly corporate videos, my client list grows—and so does my bank account. With cash now coming in from all directions, I drive down to Sunnyvale to pay back my grandmother Emeline for the quiet loans she gave me along the way—$500 here, $1,000 there—to get me through the early days when I had no paying work. I'm ashamed that I had to secretly borrow money from my grandmother. She is aging, and I am young. I'm supposed to help *her*, not the other way around.

Accepting help wasn't easy. My pride had been stitched together out of early disappointments and the unspoken promise to myself that I'd never need rescuing. When someone offered me money or food or a favor, I said thank you with a tight smile, already calculating how I would repay the debt. Gratitude and shame lived side by side in me for a long time. I bring Grandma a fresh loaf of my homemade polenta bread, the kind she loves toasted with butter alongside her morning soft-boiled egg. We sit together in her living room in front of the TV.

Now retired, Grandma sets down her crossword and lowers the volume on her *Perry Mason* reruns as a growing Emma and I fill her in on the wild life we're living in her old stomping grounds—San Francisco. She pats the top of my hand and tells us about her teenage dates, when she wore gloves and a dress.

I picture a young Grandma in the 1930s and '40s, sliding down into a basement dance floor. Afterward, she enjoys the carnival at Playland-at-the-Beach and eats the iconic It's-It ice cream sandwiches of San Francisco.

I crave my grandmother's stories and try to picture who she was as a young woman. Try to imagine myself in her shoes.

I dabble in self-reward and mild indulgences with my new money. I take Emma to our favorite outdoor mall in Corte Madera and buy her a full wardrobe at Gymboree and Baby Gap. We get lunch at the Nordstrom café, and I pick up a few new outfits for myself too. I feel completely at ease with the money I've made on my own, and I have fun with it, reveling in the autonomy and sense of accomplishment I possess.

My mother is not so supportive of my success. In groups, she lavishes praise, but in private, she tells me I should be smart, give up these silly dreams, and "get a real job," and Sal reminds me again that I'm still "hot enough" to land a rich husband. "Why do you insist on working? I can set you up with some single guys from my office," he chides and shakes his head. I ignore their advice and see them less and less. Now that I am farther away in San Francisco, I'm not running to their house to scavenge for abandoned boxes of Chinese food or old deli meats from the back of their fridge. I can support Emma and myself. I can pay my own rent and have plenty of money left to take Emma to Max's Opera Cafe for old-fashioned chocolate milkshakes for her and a vodka martini for me, traversing the city with ease.

Cliff picks up Emma for visitation once or twice a month, making the excuse that the drive is long, and he confesses that there really isn't anything to do with her. He says maybe he'll have more fun with her when she gets a little older. I see Cliff as a free babysitter, to be completely honest, and I don't hold back on the success I'm enjoying. I make no effort to go easy on Cliff when it comes to bragging. When Cliff brings Emma back from a weekend visit and one of her new Gymboree outfits is terribly stained, I tell Cliff to buy his own clothes for her to keep at his house. I'm not paying top dollar for the latest spring fashions for our daughter only for him to disrespect and ruin them. He objects, but he's not paying nearly enough child support, and this seems fair to me. I call the shots. And I'm unapologetic with my newfound power. It doesn't occur to me to have tact or to consider his feelings of being left behind while Emma and I thrive in the city. I continue growing my business, not worrying about what he's up to.

And business continues to grow exponentially. KQED asks me to caption their live programming, starting with the new weekly program *This Week in Northern California* with Belva Davis. I'm not capable of doing live captioning myself, so I headhunt real-time transcribers at the very top of their field and convince them to make room in their lucrative freelance courtroom and deposition schedules to perform live captioning for the TV programs I've booked. My business now hinges on other people coming through to produce live, top-quality work on time. The technical requirements are tricky, and the tools available in these early days of the industry are rudimentary. I have to set up each employee's home with esoteric switch boxes that hook into their landlines to send a video data signal over those lines in real time directly into the television studio control room during a live broadcast. The transcribers sit at home watching the live show on their regular televisions, typing the words into their machines. The encoded data is then sent back to the station. I regularly hold my breath for the entire thirty minutes of each show, desperately hoping the call doesn't disconnect at any point. Every single program fills me with anxiety, but it is exciting to know we're doing something groundbreaking, a feeling I begin to love and crave.

How do I convince the best in the world to work for me? I pay them more than they've ever been paid in their careers, about $200 per hour. They like the pay, and they like me. I'm a good person to work for, even though I am often many years younger than other employers. I have integrity and vision, I'm incredibly reliable, and anyone can easily see how hard I work.

At the same time, I add another lucrative service to the business: real-time transcription in universities. The idea came to me when I was asked by the Haight Ashbury Free Clinic if I could transcribe Alcoholics Anonymous meetings for a homeless man who had recently become deaf. I agreed to do it myself for free. During those sessions, I learned that a staggeringly high percentage of deaf people had only recently become hearing impaired and did not yet know American Sign Language, significantly complicating their lives. I learned that 20 percent of people in the United States are hearing impaired, and it is common for hearing loss to happen quite suddenly. Until the subject learned American Sign Language, they could not take advantage of sign language interpreters provided by schools. This gave me the idea to establish a program at universities and community colleges across the Bay Area. For five hours a day, the students at my former court reporting school could earn twenty dollars per hour by sitting next to newly deaf students in their classes, transcribing lectures in real time, and displaying those transcriptions on their screens so that the deaf students could follow along.

Sign language interpreters are in short supply for the growing number of deaf students, so my solution is eagerly welcomed by both universities and students. I charge the universities forty dollars per hour, which is half the cost of the sign language interpreter solution. With six universities on board, I am making a great deal of passive income, up to $500 per day, for organizing and sending out transcriptionists—court reporting students who still need speed practice in these formal settings to get their court reporting certificates. Because I was once one of them, they appreciate that I understand both their abilities and what they need. The deaf students receive effective accommodations, the schools save a lot of money, I make a lot of money, and the court reporters get paid quite well while earning speed practice. It's a winning solution in every way.

Somehow, despite all the obstacles, I've found the right path. At twenty-one, I was struggling to find my purpose and passion. At twenty-five, I am the master of my little universe, building a life for myself and my daughter on my own terms, and what fun it is! If I didn't so keenly remember the sweat and the struggle it took to get here, I might need pinching to believe it. Not only am I becoming an industry leader, but, more importantly, I have discovered the thirst for innovation and problem-solving that will define my career. The possibilities seem endless, and I am ready to explore them.

1978—SAN JOSE, CALIFORNIA

Yesterday was my eleventh birthday, which went by awkwardly as usual. I'm eleven and one day. For my birthday, my mom let me wear my favorite outfit—bell-bottom jeans with colorful jelly beans embroidered on the back pockets, my orange tennis shoes, and my much-loved orange T-shirt featuring a cracked and peeling decal of Lindsay Wagner in *The Bionic Woman* running with her hair down. I change into this outfit every day once I'm home from school, stripping off my sweaters, collared shirts, and corduroy pants that are too hot for these late summer days. My mom says my summer clothes are not nice enough for school. She doesn't like the jelly bean jeans, and she seems disapproving whenever I wear them around the house. I sneak them on whenever I can.

I'm aging clumsily toward my teens. I no longer wear cute dresses—I'm told we can't afford stylish clothes. We have to be practical. We can't afford conditioner for my fine, long hair anymore even though it gets tangled up in impossible knots, especially after swimming in the ocean or in chlorinated pools. One pair of shoes and one coat per school year replace my handsewn dresses and braided blonde tresses.

Mom tells me to stop asking over and over for new shoes. She says I should be grateful for the roof over my head. Sal is her husband, not my father. Sal's paying for me instead of my no-good, deadbeat dad who just up and left for England and then disappeared, avoiding his responsibility.

My brother Bryce and my dad are both gone for good. I don't know where they are.

A month or so after Daddy gave me the gold watch and left for England, Claire wrote a letter to my mother asking for help. From what I remember at the time, she said the night my dad arrived, he took Bryce from her and disappeared. Nobody ever talks about it, but I think about it every day. Where is my dad? My brother Bryce is three now, almost the same age as Giovanni. I wonder how he's doing. The grown-ups, like Grandma Emeline and my mom's brother, Uncle Jack, won't talk about it with me.

My mother has one new husband and a few extra mouths to feed. As for me, I have two sweet baby brothers to play with and care for. Giovanni is two and a half, and Matteo is just two months old. When I'm with the babies, I forget about everything else and just try to make them giggle or babble something cute.

Mom explains I should not expect Sal to treat me the same as his real kids, the babies I love and take care of every day. I'm a liability, not an asset. A living, breathing burden for my mother's husband. I don't understand this.

I understand I've been demoted. I've gone from being adored to barely tolerated, then nearly hated. They put up with me, and I should be grateful. I am not. But I don't make waves. I don't want any of Sal's attention, but my brother Evan does. Unfortunately, Evan can't help his curiosity. He's always getting into Sal's tools in the garage and leaving a mess. Craving connection and attention, Evan only gets reprimands and beatings.

I am in the sixth grade at Booksin Elementary in Willow Glen, California, a suburb of San Jose. I still haven't settled into the challenges of a new school and family. What I don't know yet is that I never will. Mom has changed so much since she got married. Did she used to talk to me more before? I guess I didn't notice until she stopped. Even though my mom is becoming someone who seems like she doesn't know me, I still love her and miss her. I still want her to be my mommy.

When I come home from school, I get started on my chores. Mom takes a break from her day and watches *The Phil Donahue Show* with a snack of strawberry Neufchâtel cheese on saltines. I vacuum the downstairs during the commercials, making perfect lines in the plush carpet. I love to watch Donahue. He looks kind with his glasses and white hair, and when he asks questions, his head tilts, reminding me of the principal at our school, Mr. Dixon. One day, our teacher got sick, and Mr. Dixon was our substitute. One of our vocabulary words that day was actually "principal"! He said there are two ways to spell it. A "principle" is something you believe in, like a value. Mr. Dixon said he is the other kind, a princi*pal*, because he is our pal. I watch the TV screen as Donahue runs among the rows of chairs with his microphone so the audience gets a turn to ask questions.

After vacuuming, I set out my homework on the kitchen counter and watch my mother as she pulls a blue card from a plastic box of flash cards. Each one features a chore like "clean the doorknobs" or "organize the laundry room." She picked up this system from a magazine like *Good Housekeeping*, and she adds more cards when she finds a job that isn't already in the box. I notice that the time it takes to keep the Rolodex organized is actually another chore. She sometimes pulls a card, makes a face, and puts the card back in for another day. Today she pulls "ficus leaves," then places the card in the back of the box for next week.

I don't mind doing the afternoon chores. It's nice in the house with just my mom and the babies, and I know she counts on me to help her with them. I get a paper grocery bag from under the kitchen sink and go into the living room to fill the bag with the ficus leaves that have fallen around their pots, careful not to disturb any of the leaves still on the branches. I know how angry Sal gets when the ficus trees are not full. I wish I could make them fuller, but every time I check on them, there are more dry leaves on the carpet. I worry about the ficus a lot.

After *The Phil Donahue Show*, my mom gets the babies up from their naps and hands baby Matteo to me. I feed him a bottle of formula while she starts dinner. I heat up the water in the microwave and carefully measure out the sticky formula powder into a plastic bottle the shape of an hourglass. I shake the bottle vigorously, but I never can get all the powder to dissolve—so I hand the bottle to my mom, who finishes shaking it and hands it back to me after squirting a little bit on her wrist to check the temperature.

While she is doing that, I peek at what she's making. I'm always curious about the menu because I love food. Mom says tonight is Joe's Special. Yay, beef! Joe's Special is a famous dish from San Francisco my mom learned from Grandma. It consists of hamburger meat, spinach, and onions. Mom stirs in an egg, then fries the concoction and serves it on toast. There is leftover cake from my birthday the night before, but the only food I don't like is cake. No one believes me, but it's true.

After I burp Matteo on my shoulder with one of his cloth diapers, I put him into his baby chair, the kind that bounces a little. I strap him in so he stays in front of the TV where I can keep an eye on him. Beside him, his older brother Giovanni is confidently sprawled sideways on a giant floor pillow. Giovanni's watching *Sesame Street*, eating dry Honey Nut Cheerios, and drinking apple juice from his sippy cup. I pull a few Cheerios from Giovanni's bowl and eat them. He smiles up at me.

My mom calls for me to get another tub of Country Crock margarine out of the extra fridge in the garage, where my brother Evan is hitting the frame of his bike with a hammer while a neighbor kid watches over his shoulder, giving him advice. Back in the kitchen, Mom is browning ground beef and has a package of frozen spinach unwrapped in a bowl to defrost. While keeping an eye on my younger two brothers, I sit at the counter and take a look at my homework, but I feel distracted. Mom is absorbed in cooking, opening cabinet doors, grabbing spice jars, and setting them next to the stove. She gets flustered at this time of day, and I feel it. Her dread rubs off on me. She must get dinner ready before Sal gets home—not too early, not too late. He likes to have dinner on the table at the right time.

I can't focus on my homework. I'm looking right at it, but I can't remember what the assignments are. Did I do them in class? Maybe I can do them tomorrow. My brother Evan comes in, and Mom shouts, "Go clean up! Make sure you haven't left a mess in the garage. Sal is almost home! Susan, set the table." Evan pivots on his heel like a breakdancer and scurries off.

A few minutes later, the front door slams. My mom, Evan, and I all freeze in place for a moment. Then we move again, trying to act normal before he sees us. I hear Sal's heavy footfalls in the entryway and his breath, resonant and ominous. He stops as he checks the mail my mother has neatly laid out on the glass-topped entry table for him to open. I hope there isn't a bill in the mail today.

Mom has all the ingredients for dinner spread out around the center island and across all the countertops. A knife for chopping the frozen spinach. A wooden spoon to stir the browned beef. The can opener and opened cans of diced tomatoes. The seasoning packet and the margarine tub. Cracked eggshells. The Wonder Bread bag is open, pieces of bread spilling out like fallen dominoes.

"The ficus trees look sick!" Sal yells to everyone and no one, making his presence known before he even enters the room. The master of the house. My mother dares not offer an explanation. I watch her from my perch at the counter, my homework still in front of me. I know I won't have a chance to do it now. I can't pay attention because Sal is sure to ask an important question I can't answer, like "Why are there fingerprints on the bathroom door?" I need to be ready for that.

Sal walks over to my mother and slaps her butt, hard. She jumps, and he laughs at her and gives her a peck on the cheek, asking, "What shit is this?"

"Joe's Special," she answers.

He sneers and walks past me to the babies, smiling at them. Giovanni runs to hug him. Evan has conveniently disappeared outside with his friend Danny.

Sal plays with the babies for a minute before going upstairs to change. My mother tells me to get my brother, so I step out on the front lawn and yell, *"Evaaannnn!"* When he appears, Mom tells him to wash up again, and although we hear water running in the bathroom, he comes out just as dirty as he was before. Inside the bathroom, the towel is covered in hand-prints, and there are pools of wet dirt around the sink.

"Susan, help me bring the dishes to the table and get your brother into the high chair," Mom instructs.

We all sit down and wait until Sal comes back downstairs. He makes everyone sit there an extra few minutes while he chooses a bottle of wine from the winery, which is actually a pool shed converted to an air-conditioned room with barrels and unlabeled bottles of homemade wine—because he's Italian. That's also why Giovanni and Matteo have Italian names, to honor Sal's North Italian heritage. They are treated like little princes in every way. My mother dotes on them, and Sal calls them *pappagallo*, the Italian word for "parrot." Sal throws them up in the air and catches them before giving them each a nip of his wine.

Being Italian is important in this house. Sal says the reason Evan and I like ketchup is because we're "Okies." This makes me think of oak trees, but Sal says it means we're from Oklahoma, which is bad. I've never been to Oklahoma. Italians don't eat ketchup except on french fries.

All of us are seated, the table is set, and the food is displayed in serving dishes on the oval glass-topped dining table. I can see my legs through the glass, my hands in my lap, waiting my turn. Sal sits at the head of the table next to my mother, the baby between them and Giovanni on the other side of him.

"I'm not going to find a shit show in the garage today, am I, Evan?" Sal says to my brother, who stares back and shakes his head no, which is a lie. Evan is like a puppy who doesn't know what will get him in trouble until it's too late. He's just nine years old, and he doesn't think about consequences yet. He knows how to avoid getting hit in the moment, but he can't think ahead. Eventually, Sal will see the garage, and then Evan is going to get it so much worse for lying.

Sal surveys the dinner spread with a dramatic pause to ensure everyone registers his displeasure. Then he says, "Pass the bread." He makes an open-faced sandwich with the toasted Wonder Bread and a generous scoop of the hamburger mixture, then serves himself an iceberg lettuce salad with Early Girl tomatoes from the potted garden and drizzles it all with Italian dressing before passing each item to my mother, who serves herself and makes an age-appropriate modification of each item for the toddler. Next, she passes the items to me and Evan before picking up Matteo and holding him in her left arm while eating with her right. Both Evan and I (and probably Mom too) want ketchup on our meals, but we don't dare get it out of the fridge.

"Do you know what we call this?" Sal addresses Evan and me. He's got another lesson he'd like us to learn. "Shit on a shingle!"

I take a moment to think about this. It doesn't sound good at all. I like "Joe's Special" better. I don't say anything, but my face is moving around, showing my true thoughts. Sal says he knows what I'm thinking and that I'm disrespectful.

We are not allowed to leave the table until everyone's finished. Mom says a family should eat together, but I think, *Maybe not this family.* Evan scrambles restlessly from the chair he's been slouching in for the past forty-five minutes before he's reminded to clear the table. He takes a few items into the kitchen, then runs off when no one is looking.

"Susan, do the dishes."

"But, Mom, Evan didn't even clear the table," I complain.

"Just do the dishes!" my mother exclaims. She has had just about enough.

Sal gets a bottle of red wine, and he and my mother go into the front yard and play with the babies on the grass, talking about their days as they do.

In the empty house, I drag the heavy vinyl-cushioned folding stool over to the sink and step up to assess the mess. The dirty pans are still on the stove, and eggshells litter the counter next to splattered beef grease, drying drips from the defrosted spinach, and an open loaf of Wonder Bread. Cooking spoons and forks, the can opener and can lids with jagged edges, tomato sauce hardening on the butcher block, oil and vinegar bottles next to the ripped package of Italian seasoning. Dirty dinner plates with crusty food waste, cups, the salad bowl, and the Italian dressing carafe on the dining table. Not to mention the baby's high chair caked in orange and brown purée and drool.

I sigh at the mountain of work, then begin with the plate of Neufchâtel cheese from my mother's afternoon snack. I scrape the dried cheese and a lone cracker into the trash under the sink and rinse that dish before getting down from the stool and placing it in the dishwasher. I continue picking up the next closest thing, rinsing it if it's visibly soiled, and stacking each piece in the dishwasher. I save the food-covered pots and pans and larger serving dishes for last because sometimes my mother does these when the dishwasher is full and she realizes they are too big and cumbersome for me to wash.

Doing dishes takes me over an hour every night, and if I don't clean them perfectly, I'm sent back to the kitchen to redo them. Meanwhile, the grown-ups watch TV or play with the babies. I watch days turn into nights while perched on my stool, staring down a never-ending queue of dirty dishes. The only bright side is that once in a while, I get to eat the leftovers from Sal's plate when he leaves a piece of fat from his steak or some meat on a chicken bone. To save money, Evan and I often get hot dogs or slices of bologna with rice or canned lima beans while Sal eats steak or chicken with baked potatoes and salad. I dream that one day, I'll be able to eat an entire steak and all the rice and potatoes I want, with real butter, not margarine.

I hear a scream and the front door slamming shut. Sal is pulling Evan by his ear while his little legs try to keep up. Sal shoves him toward the glass dining table, and Evan's head barely misses the edge as he stumbles to the floor and scrambles back up.

"Your mother told you to clear the table. Do it now, shit-for-brains!" Sal's voice is getting croaky, like a frog, and he spits when he yells. He slurps up the spit, purple stains at the sides of his mouth.

I'm conflicted. Finally, someone is telling Evan to do his chores, a relief, but I'm also angry because hitting is wrong. I never try to help Evan or say anything when this happens because then Sal will come after me. As kids, we're always just trying to get through the next twenty minutes without catching any blows, but we know they're going to happen eventually. As for me, I just try to postpone my next beating as long as I can by staying quiet and trying to do things right. It doesn't really matter how we do things because even if we do everything right, we will get hit for "the principle of the thing." That's what Sal says when he hits us but can't think of a reason at all.

Sal fills up his wine glass and goes back outside. It's a warm evening, so my parents decide to go for a swim with the babies. Evan sneaks off again, this time to his friend's house across the street. I can hear the neighborhood kids out there playing dodgeball in the street and roller-skating in front of our house when the adults come in.

"You can swim if you finish the dishes in time, Susan," my mother says lightly as she passes with her swimsuit on and the baby on her hip.

But I know I won't be swimming tonight. I scrape the leftover birthday cake into the trash and get back up on my stool.

As the punishments intensify, the adults openly discuss at the dinner table new ways to make us *behave*, whatever that means. I am grounded to my room for months at a time. The groundings start small, then compound until I'm facing three-month stints or more. From ages eleven to fifteen, I can't leave my bedroom except for school, chores, and meals. I try not to listen to the kids outside playing on the street under my window. The sound of their laughter is somehow the most difficult to ignore. Hearing the skid of bike tires or a skateboard *clunk clunking* down the sidewalk makes me long to be out there with them, but I know I have to bide my time.

Every once in a while, I'm granted a reprieve from my bedroom, and I have to take special precautions to act normal around the other kids. Otherwise, they'll be scared of me or think I'm a freak. I'm so embarrassed by what is happening in this house, so I do my part to cover it up most of the time.

I put on an act when guests come over. In truth, it's an act I need for myself. Acting like nothing is wrong is the only time I feel like a normal child. Of course, in front of other people, my parents act, too, and I always hope that, someday, the false front they present will last even when the guests retreat. It never does.

Eventually, fear becomes normal. So does being locked away, ridiculed, hit, demeaned. So does the constant yelling, dirty looks, snarls, death stares, hate. My mother's eyes always averted. Ignoring. In the room but away. Her pursed lips reveal disapproval, silence, acceptance. She knows who holds the money. She wants the money. She can justify anything that happens to us by reminding us we have food to eat and milk to drink and a roof over our heads.

But we are ungrateful, so we need to be punished.

My mother always looks nice. She smiles and laughs lightly and acts like she doesn't have a care in the world. She talks about what's for dinner and what errands she has to run and how the babies are doing in swim lessons. It's like her life is untouched by the actual things happening. I can't tell if she pretends or if she just blocks it all out of her mind.

She married a man who takes advantage of the waitress at a diner, blowing up after finishing his meal, telling her the pie crust is not crisp, his attack humiliating her. Taking extra time to draw out his criticism. We children cower in the booth, trying to send the waitress loving glances that tell her we hope she is okay and that we are sorry. We are just children, worried about this young woman who is not prepared for the hatred we are used to by now.

"Goddamn it, this pie crust is soggy!"

He stares at the young waitress. She is nervous, apologetic, standing in her uniform.

"I'm so sorry, sir. I will get you another right away."

"Now, why would I want another one when the first one is soggy? Don't you have any sense? It's the *same pie.*"

She thinks the question is rhetorical, but maybe it isn't. She waits to see.

"Never mind! Take this horseshit off my bill. I have lost my appetite."

He grunts and sneers. My mother looks scared because she can't look away and has nowhere to hide her gaze from us. I can look right into her eyes from across the booth. *She isn't really in there*, I think.

He doesn't want the pie. What he wants is a fight. He isn't wealthy and powerful yet. He's still in his thirties. He's small. He finds his power in diners ridiculing young waitresses or in his home beating his wife's young children. Soon, he'll be rich and will cast a wider net.

He's determined. I'll give him that.

CHAPTER SEVEN

OBJECTS MAY BE CLOSER THAN THEY APPEAR

"Expect nothing. Live frugally on surprise."

Alice Walker

One weekend, Cliff has Emma, so I go to a pub crawl in Alameda, just over the Bay Bridge, with my friend Yvette, another student from my court reporting school. Yvette lives in Alameda and went to high school there. I get all dressed up in a sexy silk blouse with thin straps crossing my exposed back, my best bell-bottom jeans that make my round ass look amazing, and open-toed espadrille platforms. My blonde hair is silky and long, and I wear a touch of pink gloss on my lips. I feel great. I haven't had a night out on the town in *ages!*

At one of the vibrant and tightly packed bars on the crawl route, I squeeze in between two guys at the bar trying to get the bartender's attention. I order two kamikaze shots and two beers. I'm not budging, and the men happily let me wait between them. The younger of the pair turns to rest his left arm on the back of his swivel stool and tries to strike up a conversation while I wait for my drinks.

"Hi," he says, predictably. "I'm Jake. What's your name?"

"Susan," I answer, glancing around at the bustling crowd.

"Big night out, huh? What's got you here in all this madness?" Jake asks, raising his voice slightly over the chatter.

"My friends live here in Alameda. They told me about the pub crawl," I reply, nodding toward the swarm of people packed into the bar. "It's wild— feels like the whole town showed up!"

Jake chuckles. "Pretty much. Alameda doesn't mess around when it comes to pub crawls. You get extra points for diving straight into the chaos."

I smile. "I think I underestimated it."

"Where do you live, then? Doesn't sound like you're from around here," Jake asks, his eyes narrowing with curiosity.

"I'm in San Francisco. With my daughter, Emma."

"Wait, you have a daughter?" Jake's eyebrows shoot up. "That's cool. How old is she?"

"Two. She's at that age where every day's an adventure. Keeps me on my toes for sure."

Jake laughs. "Sounds like fun. I've got a couple of nieces around that age, and they're absolutely unstoppable. It's like living with cartoon characters."

"It definitely feels like that some days," I say with a grin.

Jake leans in slightly. "So, what do you do in San Francisco? Let me guess—something creative?"

"Kind of," I say just as the bartender starts placing drinks on the counter.

Jake glances at the glasses, clearly noticing his time is running out. "Well, I work in San Francisco too. Maybe we could grab coffee sometime? Or dinner. Your choice."

"Umm, maybe!" I say, keeping it light but friendly.

From the corner of my eye, I notice Jake's friend smirking from the next stool. He looks ready to pounce with a joke.

"My pal Dan's about to give me hell for asking," Jake says with a laugh, catching my glance. "Ignore him—he's just kicking himself because he didn't talk to you first."

As I turn to go, Jake gets up off his stool and hands me a business card, saying to me, "Call me, please? I'd like to see you again." I excuse myself graciously, unmoved, but not without giving him a last good look. He is very handsome. Still, I'm here to hang out with Yvette. I'm not ready to date or pick up a guy or a boyfriend yet.

After the divorce, I am just finding out who I am. Unlike my mother or any of the TV moms of my youth, I learned already that I don't need a romantic partnership to make myself complete. I am already whole, exactly so.

I look at the business card, white with black letters. "Plan B," it reads.

Plan B? I tuck the card in my plastic Esprit handbag to be polite, but I don't think twice about him.

When I return to Yvette with the drinks, she screams in my face.

"*Oh my god!* Do you know who you were just talking to? Susan! That's *Jake Bennett!*"

Yvette is standing with a couple of friends from her high school, who all verify that I've just been talking with a legend, the handsome upperclassman who occupied all their girlish fantasies. They fill me in on the details. Jake graduated about two years before they did, was a champion swimmer, and won a scholarship to UC Berkeley. More importantly, he was an unattainable heartthrob. According to them, I'd be crazy not to pursue him.

"Yeah, okay, but no," I protest. "He gave me his card and asked me to call him, but I don't think so."

Later, at Yvette's, I argue with the girls, more than a little tipsy. They're not buying it.

"What does that even mean, Plan B?" I continue. "Does it mean he has already given up on Plan A? What kind of name is that for a business?"

I really don't like the sound of it, but I'm recently jaded by divorce.

"Or maybe he's this handsome athlete who peaked in high school and still has his pick of women," I continue. "And he hands out these cards depending on a woman's hotness. You know, so he can keep track. Does he have Plan A and Plan C cards in his pocket waiting to hand out to other girls in the bar?"

The girls all laugh, rolling their eyes.

I mimic a dude answering the phone, lowering my voice. "Hi, Jake Bennett here," I say. "Oh, I gave you my card? It's nice to hear from you. Which card did I give you? Plan B? Okay, I have a Tuesday afternoon open if you're willing to come over to my place."

The girls just shake their heads at me.

"You really don't know what you're passing up," they insist.

Am I Plan B, or is he? It doesn't really matter to me. Both scenarios end with: "No, I'm not interested."

But I do keep the card.

Yvette pesters me about it a few days later at lunch with my former classmates. "You *have* to call him! You don't understand, Susan. This is the ultimate score!"

More than a week later, I'm in my apartment alone. With nothing interesting on TV, I call the number.

Jake answers on the first ring. His joy and relief at hearing my voice is obvious through the phone.

"Wow, Susan! I didn't think you were going to call!" he says, genuinely thrilled to hear from me. Yes, he remembers me, remembers asking about my young daughter and that I live in San Francisco.

I decide to give him a chance. He asks me to dinner the coming weekend, which is perfect because Emma will be with Cliff. Jake lives across the Bay, but he works in San Francisco at the swanky Montgomery Washington Tower just up the street from me on the other end of the Financial District. He's their head valet.

Jake picks me up in a very new, spotlessly clean gray Toyota pickup truck, and he's a perfect gentleman. He opens the passenger door for me, and he takes me on a proper, grown-up date to a nice restaurant in North Beach. He's easygoing, quick to smile, nice to talk to, and completely attentive to me without being too much. He has bundles of cash in his center console, which he pulls out and leafs through when paying a parking attendant. There must be hundreds of dollars in the place where I've only seen people keep a few loose coins. He doesn't seem to care about money the way I am used to with people my age. I mean, he has enough for my rent in cash right there in the car, which gives me the sense that he's both successful and stable. He is five years older than I am, making him feel incredibly worldly and mature. I'm impressed.

Again, I note that he's ridiculously handsome. Classic chiseled jaw, thick dark hair, brilliant green eyes, a fabulous smile, an epic champion swimmer's body. I'm not really all that captivated—he seems *too* handsome. And I tell him so.

"Can I see you again next weekend?" Jake asks as he walks me to my door.

"Honestly, I'm sorry. I think you are too handsome. I'm not really interested in that. It's too much trouble. Plus, I have my daughter next weekend."

"Huh? Wow, no one has ever said that to me before."

Jake's eyes sparkle, impressed and touched—not angry, put off, or upset. He seems to be in love already.

He comes over for dinner to meet Emma and hang out with us. I enjoy our dinner, but I'm not going to sacrifice my needs or my duties as a single mother to accommodate him or pretend that I was a free spirit when I had priorities. But he seems to understand that I'm not able to go out on dates any time he feels like it, and we get to know each other slowly.

Still, I figure Jake is a popular guy and will be out with lots of different women while I'm home with my daughter, and I don't want to deal with all that just so I can show off a handsome boyfriend. But I come to find that he is also kind and sweet. He adjusts to my life circumstances and respects my parameters. He talks about his beloved group of lifelong friends and how he enjoys camping and water sports. Most importantly, he actually likes that I'm a mom, and he loves Emma right away. He doesn't flinch when I remind him we're a package deal. Instead, he reminds me that he's known about Emma from our very first conversation. He also supports my budding ambition a great deal, which is meaningful to me.

Once we begin to date regularly, we fall into a lively routine rich in variety. Emma and I accompany Jake on abalone diving camp trips, where we blend right in with his friends, mostly tech nerds, scientists, and mechanical and software engineers, some of whom are married and have kids around Emma's age.

One of the earliest recipes I perfected when I learned to cook was my signature chocolate chip cookies. I bake a batch in my tiny apartment kitchen as Jake watches football one Sunday afternoon. We're waiting for Cliff to bring Emma home from his weekend visitation. When I told Cliff about Jake, he seemed ambivalent—until he arrives at the door of the apartment with Emma, her weekend bag, and her car seat. Cliff and Jake meet for the first time and shake hands. I offer Cliff a cookie or two for the ride home, and everything seems uneventful. Jake has been nervous about meeting Cliff, so once he's gone, we talk about how it went and conclude it a success.

Around 8:00 p.m., I get a call from Cliff. He's very angry.

"Those are *my* chocolate chip cookies!" he yells into the phone. "How dare you make those for someone else?"

I try to wrap my head around this logic while also feeling defensive. It's not his right to tell me any such thing, but I try to diffuse his anger. My efforts, however, are useless.

"I don't want any other man around my daughter!" he shouts.

Cliff continues with demands about how I should behave. According to him, I need his permission to date, even now, more than a year after our divorce. I argue but try not to escalate. He is on a rant.

Before hanging up on me, Cliff pauses and lowers his voice for impact as he breathes his final words. "I am going to kill you."

I stand with the phone in my hand, frozen. It feels like he means it. It's not the "Ugh, you broke my favorite sunglasses—I'm going to kill you!" kind of threat. Or "You got invited to meet Sting—I'm going to kill you!"

It's "I am going to kill you." Click.

I'm scared.

The next day, I call my mother, and she doesn't like it either. She and Sal call Link, the lawyer who handled my divorce, and he files a temporary restraining order in San Jose, where Cliff lives. Cliff is allowed to pick up and drop off Emma, but he must not interact with me in any other way until a court hearing a few weeks later when a judge will determine if the order should be extended for six months.

When the hearing date arrives, Cliff and his family sit on one side of the courtroom. Cliff is representing himself. I have Link next to me on the plaintiff's side, along with my mother, Sal, and a couple of other family members behind me. Jake does not come with us in order not to instigate Cliff any further. Frankly, I'm sure Jake probably won't want to see me again after this.

Link makes his case, and the judge asks Cliff if he has any response.

Cliff stands and responds in a righteous tone, "I did *not* say that I was going to kill her."

After a beat, he adds, "But if I *did*, it was because I was drunk."

Well, the judge does not take this statement lightly. Disgust is plain on his face as he reprimands Cliff and makes his ruling in favor of the order.

"Mr. Mulford, the court has carefully reviewed the evidence, including your statements and testimony, and finds sufficient cause to issue a restraining order for the protection of Susan Elizabeth Mulford. While you claim that you did not make a specific threat, your statement that if you did, you were drunk only heightens the court's concern regarding your potential for harm. Threats of violence, whether explicit or implied, create a credible fear for the safety of the individual involved. Therefore, the court is compelled to ensure her protection."

The judge leans forward as he speaks, his gaze direct and his finger pointed at Cliff, who seems to cower slightly. The judge's intensity is unmistakable. Even I shrink into my seat.

"You are hereby prohibited from *any* contact with Susan Elizabeth Mulford. You are *not to approach her* in person, by phone, or through any third party. Additionally, you are to *remain at least three hundred feet* away from her residence, place of employment, and any location where she is known to frequent. Violation of any part of this order will result in immediate legal consequences, including potential arrest and further charges."

Once the judge finishes delivering the details of the order, we all leave the courtroom. My parents take me out to lunch before I return to San Francisco to pick up Emma from day care.

I give Curtis at the front gate of St. Francis Place a copy of the order in case Cliff shows up unannounced, and despite everything, I feel safe. Cliff was told in no uncertain terms *by a judge* what the new rules are, and I actually feel great about that. I don't think he needed to be so thoroughly humiliated in front of both our families, but, well, that's what happened. And maybe he did deserve it. It wasn't my fault. Of that, I'm sure.

Jake, meanwhile, does not look for a way out of our budding relationship. He stays right by my side.

1995—MILL VALLEY, CALIFORNIA

As time goes on, Jake and I grow closer, our future vibrant, but I have no interest in marriage. Why let the government have a role in our decisions any more than it already has? Why pay lawyers if we decide to break up? I certainly do not want to go through that again. Plus, I am deeply disillusioned by the idiocy of the so-called American dream, in which the definition of a successful life is marrying your soulmate, buying a suburban home, raising children, and working for thirty-plus years for a corporation to achieve financial stability and happiness and pay your mortgage. Yeah, no. Sounds like prison to me.

We consciously decide to have babies but never to marry. If we don't already have what it takes to stay together, if we can't count on a commitment based solely on our evolving individuality as unique people, a piece of paper will not give us that.

Two years after that pub crawl where we met, I stop taking birth control pills and get pregnant quickly. As my business grows, so continues my increasing reliable income, and we move from my apartment in San Francisco to a tiny three-bedroom home in Mill Valley, just over the Golden Gate Bridge.

During the summer that I'm pregnant with Jake's and my first child together, Emma is about to enter kindergarten, and I've researched the options. This house is in the district of my favorite of all the public elementary schools in the county, Edna Maguire, also known as the Garden School. Each class has two garden rows in an open field that students are responsible for tending. The teachers craft their daily lessons around the garden, relating it to every subject, be it math, science, reading, or music. Students count and plant seeds, measure spacing and depth, then watch their plants appear and compare them to their germination estimates.

I keep a townhouse in Jackson Square in San Francisco as my office and ride the Tiburon Ferry to work each day. I also set up an office in the smallest bedroom of the Mill Valley house as I await the birth of the baby. Without health insurance and with my traumatic memories of Emma's birth and the subsequent hospital stay, I'm not eager to sign up for more of the same. My friend Ramona mentions a recent story in *Parenting* on alternative birth plans. She tells me about home birth. I've never heard of it, and it sounds like a crazy hippie thing to do. I do not consider myself particularly nonconformist, but I *am* in Marin County, which is a hub for a longstanding countercultural community.

After asking around at health food stores, I find some midwives and decide to hear them out. Jake and I visit their office inside a nature-themed home with a large redwood tree growing through the middle of the house. When my overactive pregnancy bladder demands I use the restroom, I cannot find any toilet paper. Sitting on the commode, I call out to ask someone to bring me some, and the midwives calmly explain through the door that I am to use one of the cloth napkins in the basket. I reluctantly do so, wondering how many more bizarre hippie things I'll learn from these vegetarian women with their crew cuts and sustainable garb.

Back in the main room, I am ready for answers. "What do you do when the umbilical cord is wrapped around the neck?" I ask, sitting back as if I've dropped a mic, certain I've stumped them. You can't perform a C-section at home, surely.

The midwives look meaningfully at each other and exchange small smiles. They know what I'm thinking.

"We reach in and take it off," the younger one answers.

Jake and I are silent, processing the incredible logic of this answer. No immediate C-section? They've got my full attention now.

"Okay," I say. "Tell me more."

I know then that home birth is a possibility for me.

The midwives are fully trained and licensed delivery room nurses who believe home birth is a healthier, more holistic alternative for mothers and babies. We make arrangements for monthly visits to monitor my pregnancy and progress, then weekly appointments during the last month before my due date, all on a sliding scale, which is affordable for us. I learn more about nutrition and techniques for delivery that make sense. Instead of performing preemptive episiotomies, a planned incision to allow the baby's head easily through the vaginal opening, the midwives explain that they perform a perineal massage to avoid ripping. If the baby comes too fast, a woman's body will naturally tear at the appropriate spot, which allows for better healing. I want five children someday and am grateful that hospitals and the nightmares possible there are no longer requirements as we continue to grow our family.

One day, around the dinner table, Emma pats my stomach and says, "Tell the baby I said hi," like she's already practicing sisterhood. As I admire my daughter, Jake hands me a newspaper clipping announcing a new TV station based in San Francisco called CNET. This immediately piques my curiosity. The announcement is two sentences long but promises a new television concept dedicated to technology, one that seeks to make television interactive. The possibilities blow my mind.

The new CNET studio is walking distance from my office, and I pay a visit as it's being built. There, I manage to get the email address for the executive producer, Thom Bird. Since their programming claims to be tech forward, I naturally exchange emails with him about closed captioning. I convince him to meet with me so I can show how encoded text can be incorporated into the technical aspects of the network to actually create something they are currently only talking about. I already have a half-baked idea for how to connect on-screen text and full transcripts to web links and other online resources.

When we meet, I'm six months pregnant with baby Olivia, and I'm wearing a yellow cotton stirrup pant maternity onesie, my best outfit at this time. It's hard to dress with my huge belly sticking out. Thom takes one gulp of a look at me and is unable to mask his shock at seeing the woman with whom he's been exchanging intelligent emails. His opening words are candid.

"What are you? *Twelve?*"

"I'm twenty-seven," I say, then carry on with the meeting as if he has not put his foot in his mouth.

By the summer of 1995, I was already fluent in the invisible infrastructure that made live media possible. I worked with live feeds every day and was familiar with the technical demands of instant delivery speed, accuracy, connectivity, and a close to zero delay. I started thinking about how to use those same principles for the internet. There were early signs of live media being tested online, but most of it was limited to research networks. What if this kind of live information streaming could not just integrate with TV at CNET but move beyond TV to the internet itself? What if more local information could reach people globally who can't be in the room? I was standing on the edge of something entirely new, and I could feel it.

1979—SAN JOSE, CALIFORNIA

I am twelve years old, and my ulcer is acting up again.

We have health insurance, but I haven't yet been seen by a doctor for this sharp pain that fluctuates in intensity from time to time but never goes away. Like a little old man, I call it my ulcer. I hold my side and take a slow, deep breath. I know that if I relax my body and allow my thoughts to lighten, the pain in my side, which often strikes in the morning before school, will dissipate.

After the pain subsides, I bend over again to resume tying my shoes, then step over the mass of clothes strewn across the floor of my bedroom and make my way to the bathroom I share with my brother on the top floor of the house. There, I brush my teeth and run a comb roughly through my knotted hair. I turn the faucet to warm and fill an old plastic sippy cup, adding a few shakes of salt that I keep behind the mirror and gargling to alleviate my chronic sore throat.

"Maybe you need your tonsils out," my mother mentions when I tell her of my unbearable sore throats. "If you have your tonsils removed, you get all the ice cream you can eat." She never takes me to get my tonsils out, though.

I don't bother to eat breakfast, which I would have had to make myself from a cardboard cylinder of stale oats or from a box of rancid Cream of Wheat, so I pass. If I am late for school, I will get detention or be reprimanded by the homeroom teacher at my new junior high school, so I prioritize sleeping, soothing my sore throat, and leaving the house on time ahead of food.

I walk through the kitchen before leaving the house just to be seen, just so my mom or my brothers see me. I don't know why. Perhaps it's a way of telling myself I am part of something. I slip my backpack over my windbreaker and head out the door for the mile-and-a-half walk to school.

An adult could make the walk in thirty minutes, but it takes me an hour. Autumn mornings are chilly in Northern California and foggy in the pre–Silicon Valley San Jose. I feel the cold as I leave the dry, artificial heat of our house and head out, grateful to breathe fresh, natural air, which soothes my throat at the same time the abrupt cold sends stabbing pain through my neck. I pull the windbreaker and part of my sweater over my nose, creating a warm pocket from the heat of my breath.

Later, I'll stuff my windbreaker and sweater into my backpack when the warm California sun turns the cold morning into a hot Indian summer day. With my outer layers filling my backpack so full, I'll have trouble digging for my books and papers as I go from class to class.

I walk first in the wrong direction in order to round the corner of our neighborhood, making a U-turn onto the main thoroughfare. Morning rush hour is just beginning. Along the route, I see the neighborhood kids running playfully to one another's houses for carpool, excited to watch children's shows like the *The Banana Splits* and eat muffins and Pop-Tarts together before piling into a station wagon or van. I know that because when I am late, I see how the moms start their cars and turn on the heaters before the kids grab a blueberry or coffee cake muffin with a carton of chocolate milk and tumble into their seats, legs and backpacks akimbo. By the time they pass me on the road, my face and fingers are numb, but I'm almost at school.

Every day, I hope they might see me walking and offer me a ride.

"Hop in, Susan! There's always room for one more!" I imagine the pretty moms saying to me as the back door swings open and the kids make room on the bench seat. It's a dream that never comes true, but it helps to hope for it.

I see Sal drive past me sometimes too. He tries to leave the house for work before I start walking to avoid this type of awkward moment. But honestly, I think I'd rather walk than ride with him. Anyway, it doesn't matter because he never stops.

Other times, in the car with my mom and my brothers on errands or on the way to my brothers' swimming lessons, I look at people walking on the sidewalk down the same streets I take to get to school. From the perspective of the car, I know the drivers in the morning can easily recognize me. They just act like they don't see me, keeping their eyes straight ahead. Back in the neighborhood, though, I never embarrass the grownups by acting like I know they ignore me.

I tried to sneak into the gang before. I followed the other kids. *Maybe if I just show up, I can carpool too,* I thought. *Maybe they won't notice me. I'll just sneak in with the others. I won't take a muffin or a Pop-Tart, though. That would be stealing.* I acted casual, like I was supposed to be there, but the moms sent me on my way.

"We just don't have room for you," they said awkwardly.

My stupidity in trying to be included made me late for school.

I know that, by definition, one can't be in the carpool unless one's parent drives an equal amount as all the other parents. In fairness, I understand that if they allow me free rides to school, the social contract in our neighborhood will break, possibly beyond repair. Who knows what chaos would ensue?

In second period, my stomach starts to sound off. The ulcer loves when I don't eat, but hunger soon makes itself known. I dig into the side pocket of my backpack for the fifty-cent piece I stole from a big plastic tub hidden in Sal's sock drawer. He saves them for his trips to Reno, but he doesn't keep count of them. I justify that it's okay to steal from my parents so I can buy food, since technically my parents are supposed to feed me. I walk by the school office at morning break and purchase a Snickers from the snack bar. I save the change for another day and eat half the Snickers on the way to my next class. I'm glad to have made change because silver dollars and fifty-cent pieces attract attention. I'm worried that my mom or Sal will find the coin before I'm able to break it into change.

When my name is called, I explain why I haven't done my homework.

"I forgot," I say.

This produces the least amount of argument. I say it so often it becomes ordinary and kids stop turning their heads to look at me when I say it. I catch glances from classmates who were in the gifted program with me when we were younger. I can see in their eyes that they know how far I've fallen.

I don't know how to say "I didn't do my homework because Sal dragged me out of my room by my hair to beat me up. After that, all I could think about was if there was any way I could escape this place, and I eventually fell asleep. I woke up late with my ulcer acting up again."

It's so much easier just to say I forgot.

Walking home after school is better, though it is often quite warm and I sweat through my shirt. I stop at the school water fountain to hydrate before I head off. The highlight of my day, by far, is getting to the final intersection of busy streets, Curtner and Meridian, and saying hello to the crossing guard, Mrs. Wall. She's a tall, thin woman the age of my grandma, with a short, natural bob of white-gray hair under her navy-blue police-style hat. She is impeccable at her duties. With her whistle and hand-held stop sign, she commands the corner, making sure every kid gets across safely. She walks halfway into the street, holding her sign and staring down traffic. I love it when she walks all the way across the street with me. I smile gratefully at her, and I always chat with her. She asks me how my day is going in a way that lets me know she cares. Because of Mrs. Wall, I arrive home with a warm heart.

In the eighth grade, I try out for cheerleading. Sheila Shah from across the street is a cheerleader, and she coaches me, choreographing a routine for me. I'm not popular and have never been to a sports game, but I love the dancing that the pom-pom girls do when I see their routines in the quad or during assemblies.

Out of all the girls trying out, I am the very last to be announced as making the squad. When my name filters through the loudspeaker in the gym, I'm not even paying attention. I'm talking with another girl. Suddenly, Sheila slams against me in a full-body hug. She is so happy she's bursting.

"You made it!" she shouts in my ear.

What? It takes me a few beats to understand what Sheila is saying. Meanwhile, other girls begin to surround me and push me up toward the front of the gym to take my spot in the lineup of my new squad. I am really surprised that I made it since so many girls much more popular and prettier than me did not.

After the fanfare, I sneak over to the judges, who are high school cheerleaders. Still incredulous, I ask them why I made it.

"You were not the best dancer, and you are not the most experienced," one explains. "But you never stopped smiling. Your smile is so real. Most students can't remember to smile, but you beamed the entire time."

I didn't even know I was smiling, but I am profoundly overjoyed to hear this.

My mom never attends a single performance. Later, she claims I told her not to go. Maybe that's true. After I make the team, Sal says over dinner, "All the boys are going to want to get in your pants now." He laughs and snorts at his own joke and leers at me as if I am in on it.

This comment taints my win. He's gross. I know it's an inappropriate thing to say to a child.

Sometimes he goes too far. When I have bruises or when he draws blood, my mother admonishes him behind closed doors, probably thinking

Evan and I can't hear her. And she is right. We can't make out her exact words, but we know by her tone that she's putting her foot down about not leaving visible scars. She is adamant about that.

When I come in with a bandage on my knee, I see my mother's eyes dart up from the laundry she's folding on the sofa. There's a football game after school today, and I'm wearing my cheerleading uniform with all the other cheerleaders. My bandage is clearly visible through my Hanes nude pantyhose.

"What happened to your knee?" Mom asks, coming in for a closer look.

"Sal chased me up the stairs when you were at the store a few days ago. I got my bedroom door almost closed, but he pushed it, slamming it into the wall with me behind it. When the door hit my knee, my leg went through the wood, and the broken shards of wood ripped through my skin."

I explain the event like it's nothing because it is. It's so commonplace I hardly even register it anymore.

My mother stares at me without saying a word, so I go on.

"Blood started pouring out."

She still doesn't speak.

I don't get to the part where I patched up the wound with bunched-up toilet paper and tried to secure it with clear Scotch tape. Then I tied an old pair of pink jeans tightly around the bandage like a tourniquet and lay down on my foam sofa bed with my leg elevated until the bleeding stopped.

Mom tells me to unwrap the toilet paper bandage so she can see. I can tell she's already trying to downplay my story in her mind.

This is inconvenient. A visible wound.

I go to the bathroom and peel off the pantyhose under my yellow and red cheerleading outfit. Mom looks at my leg. Fat pokes through the torn flesh. She should be outraged, but she keeps calm. She cleans the area with rubbing alcohol, secures a gauze bandage with medical tape, and tells me to go to school.

When I get home after the football game, she sends me to my room.

"You lied," she sneers. "And you'll have to pay for that door out of your allowance."

I don't care about my allowance, but I care about being called a liar.

She called Sal at work, and, in defense, he told her I kicked in my bedroom door and cut my knee myself in order to frame him. She always gives him the benefit of any shred of doubt. Or, more likely, she knows he's the one who is lying but finds the truth too inconvenient to accept.

I beg my mother to come upstairs to my room. I want to show her how the rip in my skin matches the hole in the door almost perfectly.

I tell her emphatically, "I'm not talented enough to have cut my skin to match that jagged wooden hole. I can't even imagine what device could be used to cut myself like that. My knee isn't ripped clean the way a knife would rip it."

But my mother doesn't believe me. She can't believe me. If she does, she'll have to do something, and, more than anything else, she doesn't want to do something.

I get kicked off the cheerleading squad halfway through football season because I'm often grounded on game days, and I can't learn the routines because my parents say I have to come straight home from school, do my chores, and stay in my room.

At home, my mom dotes on my baby brothers Giovanni and Matteo. She reads them *Are You My Mother?* by P.D. Eastman, the story about the chick who can't find his mom.

The baby bird hatches to find his mother gone. He ventures out of their nest, looking for her. He asks everyone and everything he sees, "Are you my mother?"

I watch Matteo on my mother's lap, pointing at the pages and cuddling his head into her chest. Giovanni is snug against my mother's thigh, sucking his thumb. I wonder as I watch them from behind the sofa, *Are you my mother?*

My heart gets tight, and I feel immobilized. It is a question I cannot ask. I know the answer—she is not. I cry inside, but outside, I just watch with a melancholy, bittersweet feeling, knowing I have love to give but have nowhere to put it. No one to share it with.

I'm hopeful for a time in the future when I'll feel that love from someone else. I hold fast to this hope, although I have no way of knowing how to make it happen. I am satisfied with understanding, at least, that the love I'm worthy of is not here.

Like the baby bird, I will keep looking.

CHAPTER EIGHT

FAULT LINES

"The ultimate result of all ambition is to be happy at home."

Samuel Johnson

1995—MILL VALLEY, CALIFORNIA

O ur tiny house in Mill Valley is run-down, but it has a grassy front yard and backyard, which is rare for the hilly, wooded area of redwood groves and narrow canyon roads. Plus, there's a white picket fence in front and a redwood fence in back with a gate leading out to a kids' baseball field. Beyond that, there is a community center and a swimming pool. Emma joins the Strawberry Seals swim team. Jake, who's on the USA underwater hockey team and an expert swimmer, teaches Emma to swim and to compete. This ramshackle home—it's kind of perfect.

The house is twelve miles door to door from my office on Pacific Avenue, which is great, but the house itself is truly a wreck. Though it has three bedrooms and one bath, it's only eight hundred square feet. We're scrappy, though, so we'll make it work. The bones are good, as is the location, and it's in the Edna Maguire school district for Emma, which was our number one objective. The carpet, however, is old and disgusting, and all the rooms need paint. Still, with a very reasonable rent of $1,200 per month, I figure we can install all new carpet at our own expense and tackle all the painting. We'll make up that cost in cheap rent for the next couple of years. So we get to work. We clean it together, and Jake paints. We install new blue-gray Berber carpet in every room except the kitchen and bathroom.

The kitchen has rotted particle-board shelving and counters, so we rip them out. I buy a beautiful rustic pine hutch and free-standing armoire-style wood encasements that we can take with us when we move someday, and I also find a huge, beautiful kitchen table made out of a barn door that barely fits in the small house but will serve me for decades into the future.

Settling into the new house and waiting for the baby to come, I look forward to a new favorite show I've discovered, *Martha Stewart Living*. I watch it at the appointed time but also set the DVR to record each episode so I can watch them again and again and soak up all the small details, the recipes, the gardening tips, and the way Martha makes things look easy. Simple, comfortable, country style. I drink in all her little touches and the way she cares for her things. I learn that the smallest items in a woman's daily routine can make a remarkable difference. She reminds me of Aunt Mary Ann, who began teaching me to cook and take care of a home before my mother's second marriage spoiled that relationship.

I embrace Martha's style, and she teaches me to respect my home and everything in it. Martha always insists on quality. Her lessons reach my ears like the truth. I've found a like-minded perfectionist who enjoys the simple things—although they must be genuine. A thing doesn't have to cost a penny to be quality. It's not about the cost. It's about the value. There is no point in having something that isn't good quality. Her poise is strong and demure at the same time. In the absence of examples in my own life, I gravitate toward Martha and learn from her. I don't understand the critics who call her intimidating and demanding. On the contrary, I see her as meticulous, visionary, disciplined, and polished.

Jake rents a rototiller and prepares the soil in our side yard. We plant tomatoes, beans, squash, pumpkins, carrots, and lettuce. Mine are the best tomatoes on the planet, and they don't cost more than love, nourishment, water, soil, seeds, and sunshine. We adopt a malamute from the dog pound and name her Audrey after Audrey Hepburn, and we buy baby chicks from the farm store in Petaluma that grow into fat laying hens and live in the garden, eating earthworms and grain.

This is what I value. I make bread at home and try new recipes using the produce I grow. I find zen when I'm tending the garden, wearing my thick gloves, green Hunter wellie boots, and Jake's old Levi's 501s cut off to make shorts. I sit in the dirt with the buttons undone, my pregnant belly spilling out under my loose summer top as I weed the garden. Gardening and reading books on the weekends balance my office work during the week.

I can drive over the Golden Gate Bridge twelve miles to my office, or I can take the Tiburon ferry, a high-speed catamaran, across the bay past Alcatraz to the iconic ferry port in San Francisco and walk a few blocks to my office. I have the life dreams are made of, and we are only just beginning.

The day Olivia is born—a Thursday in October 1995—the Blue Angels are flying overhead, practicing for the upcoming Fleet Week in San Francisco. It feels like they're flying by to welcome baby Olivia, Livvy for short, to our world.

I invite anyone who wants to witness the birth to come and do so. Consequently, we have some visitors—my dear friends Marla and Ramona are here along with the midwives. Ramona brings gifts and takes photos of the laboring with a disposable camera until she's overcome by the experience. It's hot and smelly in the room. With three midwives, Jake, four-year-old Emma with the Silly Putty Ramona gave her in hand, and me, there's a lot going on, and it proves too much for Ramona. She excuses herself and vomits in our one small bathroom but returns just in time to capture the birth on film. After the photographs are developed, I see I was sweaty and red-faced, oblivious to the camera, surrounded by love. As I stare at the intimate photographs, I am aware that I've never felt more beautiful.

The entire process from first contraction to holding the baby in our arms takes about three hours. This time, I sure didn't need any Pitocin, probes, opioids, or medical ward chaos. By dawn, the midwives and my friends have gone home to sleep.

That morning, just hours after giving birth, I wake before my family, now a family of four. The house is quiet. After a shower, I am inspired to make a lasagna from scratch. I derive such joy from the serenity of the neighborhood as it is slowly bathed in first light. It's so soothing to put on a robe and make the sauce, just taking my time. It's all as peaceful and homey and unbelievably perfect as I could imagine. A million times better than *any* hospital, that's for sure.

My mom, who so wanted to be there to witness the birth, ended up missing the big event thanks to turning her phone ringer off at night, but she arrives alone around noon and, ensuring the baby is perfect, gets a few photos and is off again to get back to San Jose in time to pick up Matteo from school.

Expecting his own parents any moment, Jake takes newborn Olivia in his arms, and unbeknownst to him, I watch as he cradles her lovingly in the garden from the great paned-glass window overlooking our yard. It's a warm and sunny late summer day.

Jake is shirtless and presses the baby tenderly to his chest. She's wrapped in a thin white cotton blanket. Olivia is his first child, and he's experiencing his first private moments with our precious newborn—moments I know are irreplaceable, unforgettable, sacred. When his parents arrive, I can see the joy radiating from him—so full it goes beyond words—as he makes the introduction, our chickens clucking and scratching behind him.

"You shouldn't have the baby out here," Jake's mother, Helen, gasps as she climbs out of her car. "She'll catch a cold! You must make sure she's tightly swaddled."

Jake does not appreciate this unsolicited advice—a judgment from his mother making him feel inadequate. The visit must open some old wounds because, as soon as his parents leave, he blows up—at Emma.

At some point, Emma dropped her Silly Putty on the new carpet, and it left an oily stain. It's barely visible beneath her craft desk, a creative spot Jake built and painted for her so that she'd have a dedicated place to do her drawing and coloring. Jake does spot the stain, though, and explodes, shouting at and chastising Emma for long minutes, uninterrupted.

"Hey, let's take it easy," I call from the sofa. "The baby is new, and everything is upside down." I beckon Emma to join me on the couch with Olivia, and she cuddles up beside me. I can see the pain in her eyes, her head held down—she is internalizing his reprimands. This should be a day of limitless joy, and I want to get us all back on track. "It's okay," I coo to her. "We will not worry about this now. It has been a very confusing and emotional day."

Still furious, Jake turns on me.

"You are trying to push me out!" he shouts in front of our girls. "You will not let me be a father!"

In a shrill voice, he accuses me of making all the decisions and overriding his authority. I remain calm on the sofa, refusing to escalate. Eventually, Jake goes for a walk to cool off. I chalk his outburst up to nervous new parent emotions.

While he's gone, my friend Marla returns. She's here to assist however we need, so she whips up a blueberry milkshake with chocolate protein powder for me, then offers to take Emma out in our car so that Jake and I can be alone with the baby for some bonding time. Marla and Emma head off to the theater to enjoy a matinee screening of the Disney animated film *Pocahontas*.

I was really craving the quiet time that would allow me to get to know the new baby. Since she was born, our small house has been full of people coming and going. I've been taking care of others, answering questions, serving lasagna, scheduling visits. I haven't had any downtime since dawn, and I feel a natural, overwhelming, primal need to look into my baby's eyes and focus only on her. I especially need to take advantage of this time while Emma's away so I don't risk her feeling left out.

Olivia's newborn legs and feet slightly curve in on themselves like a ball. They're beginning to unfurl like a rose bloom as she experiences the air around her body during these first twenty-four hours of life. When her tiny, heart-shaped mouth releases from my breast, eyes rolling back in pure delirium, drunk on milk, her breath has the sweetness of a flower stamen like the hardy fuchsia I loved as a child. In my Aunt Mary Ann's garden, I'd carefully pull the stamen by the root and put the hidden end on my tongue to taste the sweet nectar—not sugar sweet but a natural syrup like my newborn Olivia's just-fed breath, as lovely as a day spent in a shade garden in summer.

I want to listen to her coo uninterrupted. I want to bask in her light.

Jake wants to use the time to attack me for not being hard enough on Emma. He's unusually agitated. *He just needs to get ahold of himself,* I think, becoming annoyed. Obviously, Emma and the new baby come first. And what about me? I just gave birth on nothing but a cup of chamomile tea.

He paces back and forth, and I feel his intensity increase. This behavior is so unlike him.

"You are letting Emma get away with things she should not be allowed to do!" he repeats. He insists I make all the decisions. He doesn't think I'm going to *let him* be a father.

I can't deal with this now, so I push him away. I'm suddenly exhausted by this, yearning to use this quiet time to nurture our baby and deepen our ties with closeness, eye contact, touch, and calm, loving voices. I don't want to take care of Jake, a grown man. I suggest he take another walk. I need this time with Olivia.

"You can be here and join me in connecting with our baby, or you can go cool yourself down," I say, intolerant. "But if you don't calm down, you need to leave the room and take this aggressive energy with you."

I'm on the bed where I gave birth less than twenty-four hours earlier. Olivia is in my arms. I don't think for one second that Jake's feelings should take priority. Until . . .

Jake leaves the room and returns minutes later with a vacant look in his eyes and a fluffy pink baby outfit in his hand.

"What are you doing?" I ask.

He does not speak. He takes hold of my wrists and begins to peel my arms away from my body. He intends to remove Olivia. I do not let go.

"What are you doing?" I repeat, growing frantic.

When I resist, he mounts the bed and presses his knees into my thighs, holding me down. Hovering over me, he pries my arms from the baby.

I should fight him off, I think, but he's much stronger than I am—and, of course, I do not want to hurt Olivia in the process. I think through these unreal choices in disbelief, but then I'm back in the moment again: If I fight him, she could be injured.

As I struggle, my legs still spread and pinned under Jake's knees, the stitches mending my perineal tear rip, leaving an open wound halfway from my vagina to my anus. Still, I do not let my arms go weak. Instinctively, I don't want to make this easy for him. Finally, though, I realize that I won't be able to stop him, and I slowly release my hold on the baby. My fear spikes. How am I in this moment releasing my hold on my newborn who is being taken from my arms?

I'm frozen on the bed, a trickle of warm blood pooling between my legs. I watch in horror as he lays Olivia on the other side of the bed and puts her into a pink outfit with a hood, which is much too big for a newborn. He doesn't know how to pick out clothes for a newborn—her arms don't even go in the arm holes, so he wraps her in it. He leaves the room and then the house with Olivia in his arms.

What do I do? What do I do? I think, already operating in shock.

This is an acute emergency, but I'm so unprepared. *I can call the police,* I think, *but what will happen?* Dialing 911 will surely escalate the situation, but do I want to escalate to that level?

If I do something now, I cannot pretend this didn't happen later.

But it is happening.

With only a second's consideration, I decide to call the police. I have no choice. I have to escalate this because my baby is in trouble. I can't fully absorb what the long-term facts of this situation may be.

Jake is outside, but in his delusional state, he forgets that Marla has taken our Jeep Wrangler to cart Emma to the movies—with the newborn car seat inside. With no car seat in the spare car, Jake paces in front of the house until he hears police sirens. Three police cars pull up and park in a triangulated formation, blocking the front of our house and half the street. Some neighbors gather on their porches to see what's happening.

The officers confront Jake, who is still holding the baby. Then two of the officers come inside to talk to me. They look so large inside our small cottage in their bulletproof vests with guns and walkie-talkies clicking at their sides. I explain what happened. Then one of the officers goes outside to fill in his colleagues negotiating with Jake.

"This baby is just a day old," they remind him.

"I have to take the baby," Jake tells them. "Susan is so controlling. She won't let me be a father."

"Where are you going to take the baby? How are you going to feed her?" they reason. "She's a newborn. Where are you going to go? Do you really think you should take a newborn away from her mother?"

He doesn't have any answers, so they talk him into handing Olivia over to them. Once an officer has hold of the baby, he brings her into the house and gives her to me.

I'm so in shock that I'm not even crying. I'm just stunned and stiff and still, overcome with sadness.

Calling 911 made this scare real. Now that the police are here, I feel safe, but my sadness turns to uncertainty: How in the world will we ever be normal again? This was a violent, insane event.

The police talk with Jake for a while longer, calming him.

"We don't want you staying in this house tonight," they tell him. "You need to find somewhere else to sleep."

With that, Jake hops into the little purple car and leaves.

The police come in to make sure I'm safe and update me on the situation with Jake. Then they leave me with a business card and tell me to call them if I need anything—or if Jake comes back.

When Marla returns with Emma, I'm still shaking. I'm in the bedroom with Olivia in my arms. Marla sees my ghostly white face and asks Emma to go into her room to play.

"Where's Jake?" Marla asks once Emma is out the door.

I can't speak. I hand her the officer's business card and burst into tears.

Marla calms me down, and I give her the highlights. She then calls the midwives, who rush right over. They examine my torn perinium, stitch me back up, counsel me, comfort me, and give me a number for a great therapist and their personal friend, Dionne, who will help me navigate what just happened. Then they bring me a microbrew beer from the fridge.

"Don't worry," they purr. "The enzymes in this hearty craft beer are good for breast milk."

Their care and the cold, tasty beer, my first in nine months, soothe me.

Around 2:00 a.m., I wake up to a sound in the living room. Olivia is next to me, but there's light trickling in through the doorway. Leaving Olivia in the bed, I walk quietly and a bit cautiously down the hallway.

Jake is on the couch, nursing a beer and watching TV. He stares at the TV and does not look at me.

I kneel at his feet and reach up for his hands in his lap. I place his beer on the coffee table and hold his hands, sitting there on the floor beneath him. I have no idea what I should do or say, but I find my mouth suddenly open, and these words come out.

"I forgive you."

Jake bows his head and begins to sob, still not able to look me in the eye. I watch his face contort as he cries.

The best thing to do in this moment is forgive him and comfort him, I think. I'm not ready to let go of the life we've been building together, the home and the love and the future. I can move past this flash of insanity. I can chalk it up to the madness of new fatherhood. I can just let it go.

If I forgive him, our family will remain whole.

Jake and I do not speak of this day ever again.

CHAPTER NINE

LAUNCH SEQUENCE

"Don't ask what the world needs. Ask what makes you come alive and go do
it. Because what the world needs is people who have come alive."

Howard Thurman

1996—SAN FRANCISCO, CALIFORNIA

O ne morning in March 1996, I'm working from home when I get a
call from Sharon, a principal at Realtime Media, the leading video
post-production house in the city. Their trendy ad agency vibe is
unlike the video game dungeon ambience of the other post-production
houses I still work with in San Francisco. I've already moved my closed
captioning encoders over to Realtime, so they know me as their least significant client. I rarely interface with the executives, except informally as
a guest at their killer client social mixers.

Sharon is only a few years older than I am but is educated, seasoned,
and chic—and she needs my help.

"Susan," she begins, hesitating a moment before finding her words.
"We have an unusual request from a high-profile client, and we don't know
how to do it. I don't know anyone else I can ask but you. The client wants a

transcript of a keynote speech and to have the text delivered on a disk to up-load to their website ten minutes after the speech. Can you do this?"

I should think it through, but I already know my answer.

"Yes, I can," I say confidently.

I definitely don't know if I can pull it off, but with enough time, I can make magic happen. I take the risk.

"Excellent!" Sharon says, but I can still hear some hesitation in her voice. "The speech is tomorrow. Keep this confidential. The client is Microsoft."

I knew Realtime Media is putting a lot of trust in me. There is no client in the world bigger than Microsoft.

An uneasy urgency travels through my body. I can't possibly arrange something like this by tomorrow. It's already 10:30 a.m. *today*. In an instant, this big break for my company has turned into a looming disaster and potentially public embarrassment.

Sharon hangs up so that she can get details from the client. You know, Microsoft. Bill Motherfucking Gates.

In a panic, I call Teri and Laura, my two best real-time stenographers for our live shows on KQED.

I say, "Bill Gates."

They say, "Yes."

And that's that.

We discuss the details into the night. I gather entrance credentials for myself and for my team, and then I make up a price out of thin air.

The next day, I sit in the front row of the auditorium alongside Teri and Laura, watching an all-star line-up: Bill Gates, Steve Jobs, Steve Case, and Jake Adams, all on stage at Microsoft's Professional Developers Conference at Moscone Center in San Francisco. There are two rows of reporters behind us watching us capture every word said on stage. The buzz travels throughout the entire media section, and journalists descend on me. I make sure they don't disturb Laura and Teri. Reporters from the Associated Press, *The Wall Street Journal*, and *The New York Times* literally crawl over their seats and crowd us during the break as if we're in a heavy metal mosh pit rather than at a civilized business conference.

"You are making a transcript? Can we get a copy?"

Business cards are thrown at me.

"Sorry, we are working for Microsoft," I say, still collecting cards.

This is interesting to me. While the reporters are busy scribbling in notebooks, I'm creating a complete transcript instantaneously. I see the demand as potential for something—I'm not sure what, but it's clearly something significant.

There's an opportunity here.

In front of five thousand attendees, Bill Gates commands the stage above us and begins his keynote, gushing proudly that the event is being broadcast over CU-SeeMe to over fifty movie theaters across the US, Canada, and international satellites. He continues by demonstrating a prototype of Windows NT with the help of a Microsoft team leader named Dave, who appears as an avatar on a screen.

I'm only half listening. Instead, I'm focused on the work we are doing. Only when I'm assured that everything is running smoothly do I turn my attention to the content of the presentation.

Bill asks, "So can we hear multiple people and talk to them all at once?"

Dave's avatar responds—but choppily. There's the sound of arctic wind blowing as he speaks. Other avatars have popped up as well. The design is unrefined, with a dystopian fantasy vibe, cartoon heads with spiky purple and orange hair floating among electric guitars in an animated desert. Bill demonstrates how the avatars can speak to one another.

"I . . . yeah, I'm backstage, connected locally . . ." one of the heads says.

It's just a demo. Microsoft is showing what might be possible in the future. Bill calls the cartoonish avatars "online travelers." The audience laughs uncomfortably at the strange video, and Bill beams with pride.

Dave goes on to explain that this technology will turn the internet from "information to communication" by allowing these online travelers to visit "virtual communities" and interact with each other by "chatting and exchanging emails."

"And so what kind of markets are you looking at here?" Bill then asks casually. "I suppose even a company can get people together and do various sessions this way."

Wait, I think, feeling outside my body, *what is happening here?* Illogically, I feel, *Is he talking about me?* Then I experience an almost ethereal clarity, a combination of doubt and elation in a singular moment. A kind of epiphany.

I glance around at the rows behind me, noting Reuters, the Associated Press, and all the major news outlets from around the world. They're busily taking notes, unaware of the discovery I'm experiencing. I peek at Teri's and Laura's screens. They're tapping away, unmoved by the words coming from the stage.

I absorb this strange reality. I'm the only one having this realization. I sit still in my front-row seat, wondering if my body is setting off a visible glow. *Did he just say what I think he said? I can do that.*

If golden is a feeling, I am golden.

The electric surge of energy inside me feels almost impossible to contain. Currently, if Microsoft or any other large multinational company wants to broadcast keynotes like this one around the world via satellite, they must commit both time and capital to flying employees from place to place to set up each participating theater. Bill may be excited to use CU-SeeMe with its stuttered images, sound cutting in and out, and demanding more bandwidth and patience than most people have. It's trying to mimic television on a network built for email. But in that moment, I know without a doubt: Text is the true currency of the internet—faster, lighter, universally accessible—and I have the tool to deliver it in real time.

As I create this new reality in my mind, I commit fact to future. I will combine text and broadcasting to create *wordcasting.* I'll generate live interactive text and stream it over this new thing called the World Wide Web, so anyone *online* can be anywhere *live.* Not from a theater—*from their home.*

In this very moment, my new company, Wordcasters, is born.

I want to jump up and run out of this conference hall to get started before anyone else comes up with my idea, but I have to finish this job. I can barely contain myself. From its first moment, my creation is already much more than an idea. I understand right then and there that this is a revolutionary global communication breakthrough.

Backstage, I oversee as Laura and Teri collaborate on the transcript in professional haste, correcting punctuation and typos while I keep the throng of curious onlookers away from them. Bill Gates walks by, in conversation with two men, and pauses a safe distance away. We make eye contact.

With his arms crossed and one finger curled under his chin in thought, I hear him say, more a directive than a question, to the man next to him, "What are they doing?" My sense is that Bill is a little annoyed that he doesn't know.

One of the men quickly approaches me. I tell him we are creating the live transcript.

I am now on Bill Gates's radar.

The next day, I call my brother Evan to let him know I am ready for that webpage he has been trying to talk me into for the past year. I finally have use for it. Evan agrees to ride his bike the fifty miles from Cupertino up to San Francisco after his shift at the Foot Locker shoe store in the Vallco Shopping Mall to meet me at my office, so I have a few hours to draw up a business plan and check out the competition, beginning with CU-SeeME.

I stock the fridge with Cokes, preorder a large delivery of Chinese food, and call Jake to tell him I have to pull an all-nighter. He brings five-month-old Olivia into the city and picks up Emma at her new day care, Kinderhaven, which is just around the corner from my office. I love that the office is actually a super swanky one-bedroom brick townhouse. With beds for the girls, a comfy sofa, a TV with cable, and a full kitchen, it's perfect for a CEO and mother of two. Emma watches her favorite VHS tape, *Wee Sing in the Big Rock Candy Mountains*, in the bedroom-slash-closed captioning studio while I work in the condo's living room.

I get to wondering—while Gates was boasting on stage about CU-SeeMe, I couldn't understand what I was missing. I read that the platform struggled with more than three to five users. How did Bill get it to work for fifty end-users in theaters around the world? I comb tech forums and user feedback reports and note many citations of instability or performance degradation under multiple simultaneous users delivering one to four frames per second, often with long pauses, out-of-sync audio, or total lock-up under load. Even Cornell University, where CU-SeeMe was developed, admitted that for group video conferencing, "performance may degrade severely" with just a few users.

On stage, Gates wasn't celebrating CU-SeeMe because it worked well —he was celebrating what it *symbolized*: "Look, video over the internet is possible." Even if it barely functioned, it was proof of concept. And this—seeing what is broken, what is missing, what no one else is solving— solidifies my initial gut instincts. I have the perfect ramp. Something people want at the highest levels in the tech world, and the solution only I have identified.

Like one of the earliest high-profile streaming failures, Mark Cuban and Broadcast.com's much-hyped Victoria's Secret Fashion Show years later in 1999, CU-SeeMe succeeded only in *doing* events—indeed, the events technically did *happen*, but they crashed out of the gate and completely failed to deliver reliable quality or scale. I know I can do better.

When my brother arrives, he immediately starts hammering away at the keys of my computer while I run back and forth, putting Emma to bed and nursing Olivia.

In addition to a webpage, we need to create a logo for Wordcasters. We agree on a large W in blue, with yellow as an accent color. Evan designs a big W with rounded corners—almost 3D—on a white background. We stare at it.

"Can you make it spin?" I ask.

"It's fine the way it is."

"I want to make it spin. Here, look." I push his hands off the keyboard and pull up another webpage with a moving animation. At this time, there are fewer than twenty-five thousand websites on the entire World Wide Web.

"Yeah, I can make it spin, but why?" Evan resists.

"I need it to spin. The W has to spin," I insist.

The two of us can be intense together, but my little brother always knows when I'm not going to drop something. His resistance is futile. Once, when we were little kids, our mom took us to a matinee at the theater and let us choose the movie. Evan was dead set on a new space movie that was popular among nerds called *Star Wars*. In that tone I was already mastering, I convinced him to instead watch something we knew would be good: a new Donnie and Marie comedy called *Goin' Coconuts*. It took FBI hostage negotiator level skills of persuasion to accomplish that. I'm not sure he ever forgave me.

"If it spins, people will know that the words move . . . in real time," I say, tossing my hair to one side after a feeding and gently laying baby Olivia's head on the soft burping pad on my shoulder.

"Okay," Evan sighs. "Get me another Coke."

He opens the graphic design program again and moves the mouse across the checkerboard screen. I put Olivia down in her travel playpen, grab Evan an ice-cold Coke, and stand behind him, giving him feedback on the design and writing some basic content before pulling up a chair. After a few hours, I watch as he adds a drop shadow to the W as it turns.

"Perfect!" I exclaim, thrilled.

On the homepage, below the spinning W, Evan adds a field that says, "Enter your email address to sign up for our newsletter!"

"But we don't have a newsletter," I protest.

"It doesn't matter," he replies.

It's 2:00 a.m. on a Tuesday morning in May of 1996. My first landing page is complete. Now to figure out how to build the product.

Soon after we launch the page, I announce the new streaming company in the Usenet groups I participate in, and I receive my first notification. The first subscriber to my nonexistent newsletter has an email address familiar to me: mcuban@audionet.net.

CHAPTER TEN

WORK, WIN, REPEAT

"Life itself is the proper binge."

Julia Child

While I build the new company and technology, Jake and I take time off once a week for long, leisurely lunches at our favorite restaurant, Tiramisu, on Belden Place in San Francisco. We love sipping icy cold French rosé and sharing cheese plates, roasted bone marrow, linguine with clams, or tagliatelle with short rib ragu in the heart of the city. It feels like the second-best thing to a trip to Paris in springtime. I soak up these moments.

We balance our home life with water sports and good food. One Friday a month, Jake and I take off work early, pick up the girls at day care, hitch up his parents' old pop-up camper, and drive up the Mendocino coast to the Ocean Cove Campground. Perched on a cliff above the sea, thirty or so friends pitch tents or park campers for three days of abalone diving, all-night campfires, and blender drinks. Our musical friends play guitar and sing rock 'n' roll or folk tunes while the rest of us throw massive logs on the bonfire, make s'mores for the kids, and pass around cartons of chocolate Häagen-Dazs drenched with Baileys Irish Cream. The pièce de résistance is always the fresh abalone specialties we prepare and share.

Unlike the rest of our group, I prefer the clear blue waters of Hawaii. I've never been one for wetsuits or freediving with a weight belt in the turbid waters of the northern California coast. While the sportier ones are out working up an appetite by freediving for abalone, three a piece, I stay back and prepare layered rustic sandwiches for when my tribe returns—a recipe I learned from the *Martha Stewart Living* TV show. Each features three kinds of deli meat and soppressata with olive tapenade, roasted red bell peppers, arugula, balsamic vinegar, and sliced provolone.

Back from their dive, the guys strip each abalone muscle from its shell and collect the meat, stuffing it inside a leg of a tied-off pair of old blue jeans. Then they slam the gastropod bodies on flat boulders around the campsite to tenderize them. Meanwhile, I prepare experimental recipes. One of my most unique contributions to the nightly abalone feasts is created by notching a whole abalone with a sharp knife and bathing it in a savory homemade sauce of apricot marmalade, salt and pepper, olive oil, and diced fresh habanero peppers. Finally, I wrap the whole abalone tightly in foil and caramelize it directly on the charred wood fire. It's a masterpiece.

I love how Jake's male friends cherish me and treat me with dignity. They are a caring, loving, respectful group of men who allow themselves a healthy dose of debauchery to prove they're indeed still boys at heart. They enjoy having fun by lovingly playing practical jokes on each other, often coltishly crude without crossing the line or hurting anyone's feelings.

One Friday, Chris, one of Jake's closest buddies, tells us about a new movie he's seen. "You've got to see it," he says, his enthusiasm raw and pure. "It's called *Jerry Maguire*, with Tom Cruise, and it's *you*. Like, exactly your relationship. I swear it's like they made the movie about you two. You're gonna love it!"

He won't shut up about how similar the characters are to Jake and me. With such a grand recommendation, Jake and I make a point of going to see it in the theater.

I have high expectations—making it easy for the movie to fall short.

I'm incensed! The female lead is a sad loser who struggles with single motherhood, and the male lead is a fabulously successful, carefree executive who takes pity on her and falls in love with her precocious, adorable child. Jake and Emma fit those roles, I admit, but does Chris see me in the Renée Zellweger role? The character, Dorothy Boyd, didn't resonate with me at all. The needy, mousy, sad sack who's cloyingly grateful for any scrap of attention and tries to level up to a fabulous life by finding a man who would, as she says, complete her? Dorothy was a concoction of a film studio. The character wasn't *allowed* to be me.

I take it up with Chris the next time I see him and straighten him out. How can there be such contrast between how Chris, our dear, earnest friend, sees me and how I see myself? I have zero tolerance for such cordial injustices. His comment, intending to be complimentary, landed sideways on me.

"I do not desire a man to complete me," I tell him. "And I will punch you in the face if you think that. No, I won't *actually* punch you in the face—but your nose will be sore for a week, and you won't know why."

Chris is stunned. I'm not sure I convince him of anything, but at least I have my say.

Is this how they see me? Needy and incapable? Waiting on Jake to swoop in and save Emma and me from a doomed life? Don't they see that I'm an *entrepreneur*, with a fantastic future ahead? *I'm* winning. Show *me* the money!

Back in my day-to-day, I have been invited to join the mayor's newly formed multimedia council. The dimly lit ad agency conference room houses a dozen executive volunteers, mostly creative media types in their mid-forties, all clad in expensively tailored clothes. They arrive in shiny silver Jaguars and Mercedes-Benzes in a city where you don't need a car. They're the type who don't need to mention that their children, all prodigies of course, attend elite private schools like Marin Country Day School, The French American International School on Oak, or Covenant of the Sacred Heart in the Flood Mansion on Broadway—a far cry from the public elementary school where Emma goes.

I missed the first meeting and have yet to figure out what San Francisco Mayor Willie Brown's multimedia task force is all about. When the facilitator warily calls the meeting to order, I realize there is no agenda. The most accomplished media execs in San Francisco look as bewildered as I do.

"We're here to give a new purpose to the historic clock tower building on 2nd Street and Bryant in South Beach," the facilitator announces with the minor confidence of a jury foreman on the first day of deliberation. I look around, waiting for clarification.

South of Market, or SoMa—what will become the home of the Giants' ballpark and the Salesforce Tower a decade from now—is, in the 1990s, a tenement area of abandoned warehouses and lofts inhabited by artists, punk bands, and gay bars. It was once my backyard when Emma and I lived at St. Francis Place, and although not far in distance from Pacific Heights where my taskforce colleagues live, it is significantly different in socioeconomic status.

What can the mayor want us to develop in that forlorn and widely shunned area in the armpit of the city?

"The mayor envisions the building as a shining marker of San Franciscan innovation—the Clock Tower Building of Multimedia," our foreman trails off, looking expectantly at us.

"A museum of technology?" one member ventures.

"No," our leader corrects him. "It's about new technology being developed here. Mayor Brown wants to establish San Francisco as the heart of the new age of multimedia."

I speak without thinking, raising my hand slightly at the same time as if I'm in school.

"What *is* multimedia?" I ask as heads turn to stare at me.

I need to know, and I think everyone else does too.

I shrug, asking, "CD-ROMs?"

The group looks at the facilitator, who looks at me looking at him, and the room collectively exhales a sigh of relief.

"That's a good question. Maybe we should define that first. What is multimedia?"

Though we break into groups and have lively discussions that night, nothing ever comes of the mayor's initiative for the clock tower building. We are, however, marking the veritable inception of the dot-com era in its birthplace, San Francisco. What we are trying to define, the internet revolution, is just about to happen.

I t isn't long before I am involved in the creation of another tech group, this one far more successful. My good friend Ramona leaves both San Francisco and *Parenting* magazine shortly after Olivia is born to take a new job in New York City for a start-up publication called *Latina* magazine. Ramona and I grew very close during our time in San Francisco, and she is devoted to my children, deeply attuned and connected to them both. Even after her move, we still spend hours on the phone on a weekly basis, maintaining our friendship despite now being transcontinental.

In one of our calls, Ramona recommends I reach out to a woman she recently met at a networking event in New York named Mary Lambert. Mary is relocating to San Francisco and has mentioned wanting to start up a satellite chapter of a networking group for women in technology. I have always been keen to follow through with good recommendations and resources at the suggestion of my wise friends, so I meet with Mary in a coworking space on Sacramento Street, where she explains the characteristics of the group she wants to build, modeled after one in Manhattan. It's called Webgrrls.

I fall in love with the idea immediately even though I don't yet consider myself a techie, per se. Mary is more connected than I am, but together, we invite a handful of women to join us for another meeting, at which we recruit our infrastructure team, a web developer and programmer, a graphic designer, a place to host a server, and an email list manager.

Webgrrls NYC holds in-person meetings and even hosts an email list eons before the advent of social media. We in the San Francisco offshoot have to create these things for ourselves from scratch. I am not a programmer or designer, so I volunteer to be the meeting coordinator. We decide to meet once a month, rotating from living room to living room. I am responsible for inviting accomplished women in technology to give inspirational talks to our group of, we hope, up to twenty or thirty women.

When Mary suddenly decides to move back to New York to work on her master's degree, I'm left to helm the group with just a handful of other women. I'm operating way outside my skill set, and I've got an infant and a kindergartner at home, a closed captioning business, and a burgeoning internet company. But I love the Webgrrls mission: to empower women in tech through workshops, mentorship, and networking and to offer new media training, community outreach, and inspiration from leading women in the industry. This is important work.

Webgrrls consumes all my extra time (what extra time?), but somehow Jake and I manage everything—both businesses, both little girls, our garden, our chickens, and our dog, Audrey. As a bonus, I'm able to beta test the Wordcasters technology by livestreaming our monthly Webgrrls meetings, which lets me work out the kinks while also gaining exposure and networking with the world's top women in technology. It's a busy time, but the steady stream of successes makes life feel full—and so satisfying.

On the other end of the spectrum, some things are not so fluid. I become a frequent service provider for Microsoft after that first huge contract of Bill Gates's keynote, and they ask me to follow him to Los Angeles. I hire six people—transcribers and tech staff—book flights, hotel rooms, and paid per diems. I put everything on my brand-new American Express card. The expenses for the group I send come to about $10,000. The job itself brings in $30,000 for ultimately just four hours of work. It's the clearest proof of concept I could ask for: a high-profile client, flawless execution, and real money.

But Microsoft doesn't pay.

For more than four months, I call their accounts payable department —first sporadically, then daily and in real desperation. I'm passed around, told the check is in the mail, stalled at every turn. Meanwhile, American

Express is demanding payment in full. My contractors are waiting to be compensated. I am sinking.

That $20,000 in expected profit—money I planned to use to grow the business—is gone. I fall behind on my own rent of $1,200, and I can't pay the office rent, our day care, or the invoices for the people who made the event possible. It sets off a cascade of financial problems that take me months to climb out of.

Later, I hear stories from others who worked with Microsoft. One of my programmers who worked on the campus told me Starbucks cut them off in Redmond after months of unpaid invoices for coffee deliveries. Microsoft employees had to leave campus to get caffeine. This was a pattern. The word in the industry was that this was intentional. It hurt small vendors like me the most.

I come to believe it reflects Bill Gates's business philosophy—cutthroat, extractive, and utterly indifferent to the ripple effects. Keeping their money in investments instead of paying small vendors on time may help Microsoft's bottom line, but it nearly destroys mine.

When Microsoft delayed payment for that job, I took it personally. Not as some clerical oversight but as a sharp reminder of how little protection I have. I am the sole provider for my small family—a young working mother with no cushion—and they are one of the most powerful companies in the world. They knew they could delay. They knew I had no leverage. That cold indifference lit a fire in me that still hasn't gone out.

Yet I don't hold a grudge against Microsoft. I can't afford to—I have to play ball. So I do something smart with that experience. When Microsoft doesn't pay, I change my payment terms across the board. From then on, it's 50 percent at booking, 50 percent before the job. Some big clients, like Hewlett Packard, push back. "That's not how we work," they say. Their accounts payable policy requires full payment *after* delivery.

I tell them, "I'm a small business. If you want me to hire an employee to chase down unpaid invoices after the fact, I'll tack on 30 percent. The rates on my sheet now are for the terms outlined in the contract. You decide."

They make some calls. My terms are approved.

This is the first time I see how standing my ground can change the rules. This is when I start learning what it means to be a *boss*.

1977—SAN JOSE, CALIFORNIA

In my new room, the small bedroom next to the stairs, is where I'll learn to measure the sound of heavy footsteps and rough, jagged breath to determine if I'll be beaten tonight. He's very good at it, almost never leaving visible bruises or drawing blood.

"What did I do?" I ask him to his face, my hair in his fist. His spankings are always some punishment for perceived wrongdoing. At first, I'm told what my crime is, usually not doing chores perfectly, then I'm given the number of spankings to expect. If I respond with something like "I don't care!" or "So what?" that number will double until I shut up. Where is Lisa Reed, with her helpful advice, now? Back at my old school with all my old friends I'll never see again. I want to ask her what I should do.

I learn on my own the best thing to say, and I keep my real thoughts in my head. Now, when I do something wrong, Sal gives me an appointment time. When that hour arrives, I leave my room and dutifully report to my parents' bedroom to receive my punishment. I sometimes think the mental anguish of waiting for the lashings is worse than being hit.

With my hands above my head and pants and underpants pulled down to my ankles, Sal forces me to bend over the bed. My face presses into the covers. Then he whips my bare bottom with his belt.

"Buckle or strap?" he asks, offering options. "You may think the buckle will hurt more, but the strap is faster and harder because air passes through the holes. So maybe the strap will hurt more. But it's your decision."

He is not my father. He is nothing like a father should be.

"Get your hands out of the way, or it will hurt more!" he commands, blaring out instructions like a PE teacher commanding me to run faster.

"If you keep telling me it doesn't hurt, I'm going to start over! Disrespectful!"

"Count out loud! Stop crying!"

The sound of the belt whizzing through the air is distinctive.

"I can't hear you counting!"

I hold in my tears until I'm back in my bedroom, and then I let them silently flow.

CHAPTER ELEVEN

PATENT PENDING

"Nothing in the world can take the place of Persistence.
Talent will not; nothing is more commonplace than unsuccessful
*   men with talent.*
Genius will not; unrewarded genius is almost a proverb.
Education alone will not; the world is full of educated derelicts.
Persistence and Determination alone are omnipotent."

Calvin Coolidge

1996—SAN FRANCISCO, CALIFORNIA

Life has a way of keeping the lessons coming. When first mapping out the technology I want to build for Wordcasters, I visit a patent attorney in San Francisco. The cool, young SoMa attorneys, who will soon set up loft-style offices and wear grunge instead of suits, are still stuck for the moment in that eighties mentality. Like the day's venture capitalists, lawyers don't yet know that their world is about to change, and most fail to understand the new economy. It's the calm before the storm.

What I built with Wordcasters was a way to stream live text over the internet in real time—before livestreaming existed. At the time, nothing like it existed. My team and I were inventing a new form of digital communication. The patent I wanted would cover the method and system that made this possible, from the transcription interface to the way the text was delivered online. I meet with a team of three lawyers who look like the kind of guys who don't know much about technology. They hesitate and stutter when they see me. To them, I am a young girl looking for a technology patent. When I explain my process to them, they have no idea what I am building or why. They can't see the future I can see. Even so, they explain that I can indeed patent my invention. Then they tell me the price.

Ten thousand US dollars.

Okay, hold my beer.

Now that I know how much it costs, I go about finishing the product and raising money while also working with more clients to get enough of a cushion to afford the patent. It doesn't take me very long to earn ten thousand dollars, but with no real urgent need to get the patent—or so I thought—it takes me a year to return for it.

In the late nineties, we are inventing in real time. The internet is still dial-up and clunky, but we push it to do things it has never done before. I'm not just streaming words—I'm building a system that allows live events to be experienced remotely and interactively. I am experimenting with artificial voice widgets that Microsoft is developing and testing a genie they created that comes out of a bottle on your screen and reads the live text back to the user. Other widgets I want to develop could be fashioned to read the text on the user's desktop in a voice of their choice by training the plug-ins to learn the voice of the actual speaker or a celebrity voice like Arnold Schwarzenegger, just for fun. Every week brings a new breakthrough—new features, interface improvements, impossible things made possible. We aren't copying anyone. We are making it up as we go, and somehow, it works.

After we build a unique system, one that solves a new problem—how to make live events digitally inclusive when video bandwidth is still too limited—I return to the lawyers' office to patent the technology. Now, the investors are asking me to produce a patent before I begin serious pitching to venture capitalists.

"You waited too long," they say.

What? My life flashes before my eyes. Turns out, securing a patent is only possible within one year after announcing an invention to the public. I was featured in media articles—and talked about my tech—more than a year ago. Add to that the fact that patent applicants are barred from earning revenue from their product, and I'm cooked. I invented the first live text streaming platform and pioneered interactive event broadcasting on the early web. I've been performing work with the tech I built. I secured major clients (Microsoft, HP, Sun) based on the product's functionality. I'm devastated.

I want to kick myself! I never asked if there was a time limit. How could I have known? *Now I'm dead in the water*, I think. The patent would protect our rights to the invention and discourage competition. It's supposed to be the safety net for my company, and now the net has evaporated!

When I get over my shock, I look around the wreckage of my business. What a tragic error.

But then I start to redefine my thoughts.

Okay, this is a major blow. Some would say fatal. But it can't be the end of this road. What can I do to fix this? What do I still have of value here?

I twist my mind inside out to find some way forward with what we *do* have, patent or not.

How can I protect my invention from competitors?

I come up with a new pitch, adopting an idea I've heard other technologists call "open source" technology. I pay attention to the movement against patents, which, some argue, actually inhibit forward progress. It's an intriguing idea—and one I can spin to suit me. *I never wanted a patent to begin with*, I tell myself—until it becomes true.

"We don't need a patent," I announce to anyone who questions the fact. "We're moving too fast."

"But what about competitors?"

"Competitors can only copy what I've done so far," I confidently tell investors as if I've known it all along. "They can't copy what I'm doing *next*. I don't have time for all the lawyers and waiting periods. We're innovating so fast—we are building so fast. They'll never catch us because they can't think like I do."

Not having a patent actually sets us apart. Who needs a patent when we have my brain? No one can copy *that*.

1982—SAN JOSE, CALIFORNIA

rack! The back of my head hits the wall. My hair is in Sal's fist, and his pockmarked face is so close I can see every pore. He forces me against the wall and holds me there, one of his knees pressed against my legs. Spittle flies through his purple and blue lips, and he scowls. I'm no good, he tells me. I have no respect.

I'm careful to avoid eye contact since it will only earn me an extra whack on the side of my head. The skull is a perfect target because my hair will hide any bruises. Without moving my head or body, I let my eyes dart toward my mother, who's sitting on the sofa just a few feet away. I can see the back of her head. With all my heart, I will her to turn around and see me. See this. As if she isn't already fully aware of the violence going on behind her. Still, with my head held against the wall and her husband screaming at me, I long for her to put down her magazine and save me. To finally tell this brute she chose as her husband to stop.

Instead, she licks a finger and turns the page of her magazine. A feeling of loss cascades down and through me as the curls of her pretty brown hair softly shift across her shoulders when she turns her head to the left to read the next page, as if she's alone in the room. She's only a few steps from us, but it's like she is far, far away.

Finally, my attacker's interest wanes—maybe because I won't fight back, a bore, or maybe because his glass is empty. Either way, he lets me go, hurling me toward the floor with a few final insults and stumbling barefoot out of the room, bunions twisting his toes into pretzels. In his stretched-out white undershirt, black leather belt undone like a threat, he makes his way to the kitchen to fill his glass with more wine.

"You're so stupid that your dad doesn't even want you," he yells at me as I start up the stairs to my room. "And tell him he owes me a lot of money for you and your shit-for-brains, pansy-ass brother!"

I'm already in the hall at the top of the stairs when I hear him add, "Oh, yeah, that's right. You *can't* tell your dad because you don't even know where he is!" He cackles at his own joke as I quietly close my bedroom door, hoping this is it for the night.

When things quiet down, I sneak downstairs and swipe a pair of scissors from the kitchen counter. Then, in the bathroom, I cut off my hair, shearing it as short as I can, though leaving my bangs intact. It looks so good.

Then I slip back into my bedroom undetected and listen to KFRC AM on my transistor radio while thumbing through one of Mom's old issues of *Reader's Digest*. After a while, I hear Sal's heavy breath coming up the stairs. I turn off the radio and hold my own breath, lying still in the dark. Underneath the covers, I'm still wearing all my clothes. The creak of the floorboards tells me Sal has passed my room, and my body slowly releases its fear.

But my relief opens space for a different kind of pain. If Sal passes my room, it means he's headed for my brother's. Soon, the screaming will begin, a sound more painful to me than getting hit myself. When I'm on the receiving end of Sal's blows, I don't scream, ever. My brother always does. I try to ignore the sounds. *Just fall asleep*, I tell myself. I do not turn on my light.

CHAPTER TWELVE

A QUIET WARNING

"It is impossible to have your eyes open and not have your heart broken."

Dr. Gabor Maté

"**M**ommy, can we go to Hearts and Flowers today?"

Five-year-old Emma looks up at me, a large bar of green sidewalk chalk in her hand and a Giants baseball hat askew on her head. We both love Hearts and Flowers, a Victorian-style flower shop and café in the tony bayside village of Tiburon, a few country blocks from our house on East Strawberry Drive in Mill Valley. Beautiful peonies and roses with massive heads in all colors pour from huge, galvanized silver buckets so plentiful you can hardly pass without being brushed by petals. Their buttery, lemony madeleine cookies, Emma's favorite, cost a whole dollar each, but they're deliciously worth it. On our regular visits, we browse the stacks of cut-out paper doll books and ogle the pretty English teacups on display.

"No, not today, honey. Mommy has to work."

"I'll take you out for a bike ride a little later, okay?" Jake chimes in.

Emma seems reluctant for a moment as she sits down at her periwinkle blue drawing desk in the family room, clearly wanting Mommy time.

I have a Webgrrls event at the Sir Francis Drake Hotel next week. Almost overnight, our little group outgrew its living rooms. From starting at fewer than twenty tech women, we've now multiplied to several hundred members, and we're running out of spaces big enough to hold us all. When we have big-name speakers, the historic San Francisco hotel offers us their ballroom at no charge. This month's speaker is the extremely popular Brenda Laurel, who is launching a highly anticipated gaming company for girls called Purple Moon. I tell Emma that's what I'm working on and that I will take her with me to the event. She isn't satisfied, but she busies herself with her drawing.

Suddenly, Emma pops up from her desk, raises one tiny hand, and points a finger upward, indicating she has had an epiphany.

"I know! When I grow up, the *daddy's* gonna work!"

With that, she sits back down, a smile of satisfaction painted across her cherub face, as if she alone has just imagined a new world in which mommies stay home and daddies go to work. Imagine that! Jake's eyes meet mine, both of us laughing silently, while I wonder with a pang of guilt if I'm working too much.

With Emma's kindergarten year underway, our garden thriving, baby Emma growing sturdy as a lumberjack, and our chickens laying regularly, life is whole. The vegetable beds overflow with vines and green shoots. Our home in Mill Valley, tucked beneath a few trees and dewy garden mornings, is a small sanctuary.

But even in that peaceful setting, there are moments that begin to unsettle me.

I notice little things—stray comments, odd silences, sudden changes in mood. A shadow passes over Emma's face at unexpected times. She clings to me for longer than usual after returning from her father's house. Things a busy mother can easily overlook. And I try to.

At first, I convince myself not to overreact. I chalk it up to normal growing pains and the struggle of growing up between two homes. I want so badly to believe everything is fine, so I tuck my concerns into the corners of my mind. They don't mean anything, I'm sure. Love and stability will smooth everything over.

But the signs don't go away. Instead, they gather quietly in the in-between spaces—in bedtime stories interrupted by unexpected questions, in whispered worries and guilt that don't belong to a five-year-old's world. Themes start to repeat: worries about boundaries, about safety, about needing permission for things that should never be in question. They thread themselves through our ordinary days like tiny warning bells.

I've worked so hard to build a home full of real safety and joy, not just the appearance of it. As these changes unfurl, I stay watchful because I know how important watching is. All the while, though, every part of me hopes, aches, to be wrong.

Jake notices too. One afternoon, while we are all piled in the living room, Jake and I lock eyes. Without a word, we both understand.

This is real danger.

Something is happening, and it's bigger than either of us wants to admit. My intuition, the voice I've been trying so hard to quiet, is now screaming at me. I can no longer pretend it's nothing.

A part of me breaks—the part that still believes I can will our life into something untouched, something safe, just by loving and working hard enough to earn the carefree independence we deserve. I have to step out of hope and into something more challenging. I have to face what's real: that, while I may be safe and free, my daughter may not be.

This understanding leads me to seek counselling. I make an appointment with the county's child advocacy office. The day I take Emma in for her appointment, I'm surprised to learn that I won't be allowed in the room with her.

"Don't worry," the specialist says, hushing my concerns. "Emma will color and play with dolls, and we'll observe her. When appropriate, we'll use gentle conversation to probe for any signs of stress or unhealthy experiences."

I sit in the waiting room, trying to focus on the sound of the clock on the wall but mostly wondering what's happening behind the closed door.

After about twenty minutes, the counselor emerges with Emma, who's smiling politely but pensively and holding a picture she colored. The specialist kneels down, thanks Emma for coming, and tells me that the next step will be to speak with her father. After that, they'll make their recommendations.

In the 1990s, child protective practices were evolving. Interviews were less standardized, but key elements still guided how child advocates, social workers, or investigators approach these situations.

When the results finally come in, I open the envelope at the kitchen table.

"Inconclusive," the report says. "Assisting the child requires documented or credible evidence of risk."

I read the report carefully, looking for something more, but there's no certainty to hold on to. I call the specialist's office to ask for clarity, but the answers I receive are just as vague.

"Keep track of what Emma says," the counselor tells me. "If anything changes or intensifies, contact us again."

Without a defined path forward, I tell myself this must mean there's no clear cause for concern. Surely, the professionals would know if something were truly wrong. I turn my attention back to our everyday life, still listening closely but trying to stay neutral. And waiting, in case the next step ever becomes clear.

1979—SAN JOSE, CALIFORNIA

One day in summer when I'm not grounded, I'm allowed to go to the community pool in Brighton Square with some girls from the neighborhood. Our family does not belong to the community pool because we have a pool in our backyard, so we don't pay monthly for an additional pool membership. But I really want to be part of the community pool's kids group because they can walk to the pool by themselves and just play. There's a lifeguard, and it looks so fun.

I imagine they have days like today every day. All the kids are playing and screaming and splashing around and just having a great time, and here I am right in the middle of it all. It is fun to be at another playground with all the kids from our neighborhood, but I know somehow I don't really fit in. My house is the one house on our street the kids don't visit. The one front door that kids don't knock on to see if I can play. I'm splashing around in wonder at my luck for being here, learning how to act normal by watching my peers

When I get out of the pool to dry off and warm up in the sun, my observant friend Diana nudges me. "Susan," she says, "who is that man behind the fence? He's taking pictures of you."

I immediately get a weird feeling in my stomach. Creepy! The thought of being spied on is at once scary and mysterious, like I'm in a Nancy Drew mystery or *Harriet the Spy*. It can't be real. I walk a couple of steps to the chain-link fence and see a man with a camera at his eye, snapping away.

I recognize him.

"Baba?" I call out to the man, confused.

I can sense Diana nervously eyeing me from the deck chairs.

"That's my grandfather!" I call to her.

Then I turn back to the furtive photographer.

"Baba! What are you doing out there?"

But Baba just takes a few more pictures and walks away, fast.

"Baba! Wait!" I shout after him.

I want to ask Baba where my dad is. I want to ask him anything, just talk to him, but he gets in his Cadillac and drives away. I can't run after him because I lack the plastic wrist tag that will allow me reentry, and anyway, I'm dripping wet in my swimsuit. I watch his car disappear into the distance.

Sometimes, I question the reality of events. Was that Baba? What was he doing? Why didn't he talk to me?

Years later, it hits me. Baba was probably taking pictures of me so my dad could see how I was doing. Maybe he did it more than once. Maybe, unbeknownst to me, he followed me, secretly documenting my life for my father. For now, though, I put it out of my mind. I don't see Baba again for another five years. In the meantime, I keep the sighting to myself and the kids at the pool who saw him. There's no grown-up I can ask about it, anyway.

CHAPTER THIRTEEN

WHITEOUT CONDITIONS

*"Grace, my friends, demands nothing from us but that we shall await it
with confidence and acknowledge it in gratitude."*

Isak Dinesen

1996—MILL VALLEY, CALIFORNIA

By Christmas 1996, Wordcasters has taken off. I'm featured in newspapers and on radio shows. I recently bought my first new car from the dealership—and it was easy! The salesman didn't object to my income and never reappeared to deliver bad news. I could afford to buy a brand-new Jeep Cherokee SUV in a shimmery gray taupe color that I loved. What a thrill it was. I've officially achieved a new level of success—on my own—that I didn't see coming, and I am grateful to the universe.

With more newspaper and magazine articles on their way, I am asked for a press kit and headshot.

"Sorry, what's a headshot?" I ask the requesting reporter.

When he explains, I find a professional photographer and sit for one. It's agony. Proof sheets show me in various awkward poses, turning seriously into the camera with one crooked finger under my chin for no reason or standing with my arms crossed in front of me, so fake it makes me want to gag. *This is so stupid*, I think. I wonder if, when the photos are published, people will be able to tell how phony I feel in them. I can barely choose one photo I like from hundreds.

Now that I have headshots, I also have to do something about my name. I'm still legally Susan Mulford, stuck with my ex-husband's last name. It kills me seeing printed praise for Susan Mulford, as if Cliff deserves any of the credit. I know I need to change my name, but to what?

My birth name, Dorsey, is a relic from the past. Sal so often used that name as a derogatory term that I can't stomach returning to it. "You're a Dorsey," one of Sal's deepest insults, still rings in my ear. Plus, since my father did abandon me in childhood, I feel no obligation to defend the family name.

I never want to marry again, so I'll carry any name I choose for the rest of my life. My mother's maiden name was Bush. That doesn't appeal to me. My grandmother's maiden name was Wilson. Susie Wilson was an acquaintance of my parents. Further back, my great-grandmother's maiden name was Hicks. No, that doesn't feel right. In short, I exhaust the family tree and come up broke.

So I decide to simplify the issue. One Sunday afternoon, after vowing to choose a name and print new business cards on Monday, I sit on our living room floor with the San Francisco phonebook. With Olivia crawling around among her toys and Jake supporting me from a nearby chair, I close my eyes, open the phone book randomly, and put my finger on a page to discover my new name. The strategy seems reasonably random and uncontrived. I'm aiming for a name that's easy to spell, that's common but not too common, and that suits me.

Excitement fills me as I plunge my index finger down onto a page.

"What is it? What is it?" Jake asks.

I open my eyes, and my face falls.

"Chan."

"Oh, no, that doesn't suit you."

"No."

I try again.

Jake picks up Olivia and bounces her on his knee.

I open my eyes and chuckle.

"What is it?" Jake asks.

"Chan again."

"Okay," Jake says. "Flip to the back of the book."

"One more time," I laugh. "Or however many times it takes."

My hope of landing on the perfect name on the first try has given way to practicality. I'll do this until I get one.

On the third try, I open my eyes.

"Quinn."

I look up at Jake. He nods, considering it.

"Susan Quinn. Yeah, I can see that," he says after a beat.

I roll the name over in my mind a few times and glance around the room. I'm probably part Irish, so, yeah, Quinn works. It's short, rather common, and cool too. I like the letter Q.

Then I notice what's playing on TV in the background and point, mouth agape, to the television set. Jake turns to look.

It's a rerun of *Doctor Quinn, Medicine Woman*. There it is—there is the magic!

Jake and I laugh and high-five.

The next day, I print business cards with the name Susan Quinn on them. Then, I head to the DMV and let them know I've changed my name. Wildly, they don't ask for any documentation. I simply take a new photo and pay ten dollars, and my new driver's license arrives in the mail. I'm officially Susan Quinn.

When I tell my parents about the change, they shake their heads. I'm sure they'll have a lot to say about it behind my back, but I don't care. I have a name that's all mine and not a representation of any of the dead-beat or control freak men who dominated my life until now.

Jake and I decide to take a real vacation from Christmas to New Year's. Tech basically closes down in December, and Evan's girlfriend's family has a small condo in Sun Valley, Idaho, that we can rent for two weeks at a family discount, basically the cleaning fee. It's a no-brainer. We drive to Idaho in our new Jeep Grand Cherokee filled with presents for the girls, our box of Christmas ornaments, a star-shaped tree topper I made out of tin foil, the homemade string of popcorn I use every year, and our winter clothes. We listen to the books on cassette tapes I bought from Borders, and we take turns at the wheel and drive through the night so that we can avoid wasting money on a hotel.

We pass through Winnemucca, Nevada, around 3:00 in the morning. Whoopi Goldberg is reading *How Stella Got Her Groove Back*. When we stop for gas, I notice the concrete is dangerously icy. I almost slip and fall several times as I cross the parking lot to the bathroom. It's also beginning to snow, but the ice is my main concern. I tell Jake about it and note that we should probably drive carefully.

Jake ignores the warning, reminding me we are on a schedule to arrive in Idaho so we have a full day to stock the fridge and pick up a Christmas tree when we arrive. Back on the highway, he hits ninety miles per hour, which sparks an argument about safety. He's annoyed by my micromanaging and scolds me for my backseat driving. I tuck my legs up under me and shut my mouth. I said what I had to say. He can drive how he wants.

The audiobook plays on. The girls are asleep in the back seat, all bundled up in blankets. The fog and snow intensify as we plow ahead.

In the thick fog, we come up fast on a car that's crawling along, cognizant of the conditions. Jake swerves into the left lane, screaming as he does. His hard left sends our Jeep into a quick counterclockwise spin.

We spin and spin, careening from the left lane back into the right.

"Oh, no! Oh, no!" Jake cries. He turns the wheel frantically, hopelessly.

Like a whirling ice skater fixing her eyes on one spot for balance, I eye the side of the highway.

"It's okay. It's okay. It's okay," I calmly repeat.

I can see as we are spiraling that we're heading toward a bank of fresh snow, five feet high. Not toward a cliff or a ravine. Everything will be fine.

When we make impact, however, it's from the rear, and our landing isn't a soft one. First, we hit a large highway lamppost on the driver's side. It falls backward, hard as a redwood tree. That first impact propels the car forward again, and we crash into a speed limit sign before finally coming to a stop. We've hit the second pole with such force that the front corner of the Jeep on the passenger side is crushed in on itself. Good thing I had my legs tucked underneath me or else they'd be crushed too.

We're alone in the dark on the highway. We sit, for the moment, in silence.

Jake and I are alive. I turn and check on the girls. They've slept through the whole thing. I wake them up and check them for injuries. Miraculously, none of us has even a bruise!

Jake pulls himself out of the truck, crying uncontrollably and chastising himself. I follow him out to console him. We're covered in small pieces of glass, courtesy of the broken windows. Out in the snow, I am stunned to see that the car is completely mangled. My new Jeep is totaled. Even the tires are no longer round.

For now, I ignore the scene and wrap my arms around Jake.

"We're fine," I remind him, my voice soothing. "It's already over, and we made it."

He needs to calm down. It's 3:00 a.m. and freezing cold, and we are in the middle of nowhere with no shelter.

A Suburban pulls up alongside us, and a family peers out.

"Are you okay?" a passenger asks, rolling down her window.

"Yes!" we shout back, still in shock.

My mind is racing. Should we leave the car with everything inside? Our gifts, clothes, everything?

As the Suburban disappears into the deep fog, I take a moment to think.

"It's getting cold," I finally say to Jake. "We'd better stop the next passing car and get out of here!"

Jake is still crying. "I could have killed us," he says through his tears.

"But we're all okay," I reassure him again. "Let's find shelter."

We get back in the Jeep with the girls, even though all the windows are broken and the snow is coming in. I wrap Olivia in blankets and hold her on my lap. Emma snuggles up beside me.

The next vehicle to pass is a big white Ford, a four-door rancher's truck with a white-bearded man in a cowboy hat at the wheel. We all pile into the warm and cozy cab of his truck with a couple of our bags and toiletries. I'm in the back with the girls, and Jake takes the passenger seat. The man ferries us to the Holiday Inn an exit away.

In the back seat of the truck, Emma presses against me.

"Is he Santa Claus?" she whispers into my ear, her eyes filled with excitement. I know she slept through the accident and don't want her to absorb our worries. Jake is chatting with the driver and in good spirits knowing we are rescued. This is what a good parenting team does. I'm focused on the girls.

I eye the driver in the rearview mirror and smile—Emma is watching my reaction. With our driver's cheeks pink from the cold night air and his full white beard, there certainly is a resemblance.

"*Could* he be Santa Claus?" I whisper back, widening my eyes to elevate the magic of her discovery.

I can see Emma pondering my reply. Then, with a twinkle in her sleepy eyes, she smiles and nods.

"I think so too!" I say, pulling her in close.

From the Holiday Inn lobby, Jake and I phone the state troopers, then make our way to a room, where we find a welcome bit of sleep. At sunrise, I brush my teeth, not realizing my toothbrush is sprinkled with tiny glass shards until I feel a slice to my gums. It's a good lesson, though. We'll have to check all our clothes, Livvy's diapers, everything for glass dust.

One thing at a time, I think. We will get through this. We are alive and warm in a hotel.

Once we've showered and grabbed some breakfast in the hotel restaurant, a state trooper picks us up and takes us to the wreck.

"I drove past the crash site on my way in," he says as we drive. "Are you sure you folks are the ones from that car?"

"Yes, we're sure," Jake and I reply. Obviously we're sure.

"By the sight of it, I didn't think there'd be any survivors."

The Jeep looks even worse in the daylight. We get all our things out of it—our Christmas packages, clothes, toys, everything. Nothing was stolen in the night. The officer then delivers us back to the hotel and helps us arrange for a tow to a nearby junkyard.

Back at the hotel, we have to decide what to do. We rent a minivan in town, but we're not sure whether we should drive back home to California, which seems like the automatic answer, or continue on to our little condo in Idaho. If we go back, I reason, this will be a tragedy we always remember. There's not much for us to do at home but lick our wounds. But if we go on, we can relax and enjoy the holidays. We can always handle the car situation when we're back home again, two weeks from now.

Just like that, we decide to continue to Sun Valley. We arrive a day later than expected, on Christmas Eve, so we immediately head out in search of a Christmas tree. Unfortunately, everything in town is closed except for a small grocery store, where we buy some vegetables, a whole chicken, and other things to cook for Christmas dinner.

The town nursery is closed, but as we drive by, we spy a lone Christmas tree wrapped in mesh and leaning against the fence outside the front door. We assume it was left there for us (thanks, universe!), so we take it back to the condo and decorate it. Someone surely left it out for just this type of unexpected emergency. Who buys a Christmas tree the day *after* Christmas?

With the tree properly festooned, I roast the chicken in the oven, and we watch Christmas movies on TV and enjoy a very cozy vacation.

In Idaho, we notice solid, snow-safe Jeeps all over town, so we decide to buy a used one, a Jeep Wagoneer, which we score for $7,500 cash. It's black and has wood panels on its sides. After our two weeks, we pack up the used Jeep and drive the minivan back to Winnemucca to return it. Jake and I are both skittish and afraid of driving on ice after the trauma of the crash, but we manage. We pass our wrecked car in the junkyard on the way out of town and continue on home.

When we arrive back in Mill Valley, we are relieved to be home and also happy that, after those two weeks, the tragedy we experienced already seems like the distant past.

We don't expect to find a much larger tragedy brewing.

As we pull into our driveway in the dark, I see a manila envelope on the doorstep. I fetch it and bring it inside as Jake unpacks the car. Once the girls are safe in their beds, I open the envelope.

Inside, I find legal papers—page after page of accusations, typed and with legal stamps and filing titles. It's a lawsuit, I realize, containing numerous citations claiming that I'm an unfit mother. In total, there are sixteen counts of child endangerment and a demand for full custody of the "minor child, Emeline Lucia Mulford."

My mind is alive with fire—and then dark.

When I took Emma to be evaluated for possible abuse by a child counselor, I knew there would be consequences. As predicted, Cliff was irate. He called me, screaming, and demanded to know what exactly I was accusing him of.

"I'm only responding to the things Emma has been saying, asking me for boundaries. I'm worried about her," I explained.

I never accused Cliff of anything publicly or privately. I never saw a lawyer about my concerns. A lawsuit never crossed my mind. I was simply concerned about Emma carrying things no child should have to carry, and I felt it was my utmost duty to advocate for her. I never considered that doing so would put her at even greater risk or that there was any possibility that she could be taken from me.

Cliff's mother was a family counselor. I assumed that meant he would understand my actions and the process that followed. As Emma's mother, I had an obligation to make sure she was okay.

Instead, he's suing me for custody.

Unbeknownst to me, after Child Protective Services interviewed him, Cliff hired an attorney, a woman determined to make a name for herself. The two conspired to declare me an unfit mother, claiming I'd brainwashed my child against her father.

They say I alienated my daughter from her loving father. The paper suggests I should not be allowed visitation unless under supervision and that I should be required to undergo a full psychological fitness assessment.

At first, I'm entirely shocked. Disbelief, then panic and fear wash over me in rapid succession. Being accused in this manner is something I never expected. I'm completely blindsided.

I go through the documents a second time, slowly reading each accusation and trying to understand them. I have to read the words again and again. What are these words saying? What do they mean? Where did they come from? What can I do? How much danger are we in? It is clear to me that we are being attacked. The only thing to do now is to wait for the court to see what is obvious to the rest of us. No one could possibly believe these lies, certainly not the professionals in the justice system.

Yes, I tell myself, *the professionals will know exactly what to do.*

1983—SAN JOSE, CALIFORNIA

I'm fourteen when my mother and grandmother decide to take a trip to France. The night before they leave, Sal beats me up again. He's drunk and slow, and he falls down while holding on to my clothes, ripping the knitted vest I'm wearing over my shirt. When he falls, I scramble up and slip through his grasp and race out the front door into the dark night. Typically, if I make it to the door, he doesn't follow, not wanting to be witnessed by the neighbors.

It's always a split-second decision to either take the beating or try to get away. If I make it outside, the door will be locked behind me, and I'll have to sleep in the garage for a few hours until Sal unlocks it. This time, as usual, I hear him lock the door behind me, so I run all the way to my friend Elisabeth's house. I can reach her window from the street, and she will open it to let me in without her parents knowing.

This time, I know I have to run away for good.

The next day, I wait until the house is empty, my mother having left for the airport and Sal at work. In fear of getting caught, I collect some of my things, filling black plastic garbage bags with everything I think I'll need for my new life on the road: clothes for unpredictable weather situations, a couple of books, a sleeping bag, a pillow, some magazines, and as many snacks as I can fit, like granola bars and a half jar of peanut butter.

I've seen episodes of ABC's *Afterschool Specials* about runaways, and I imagine being homeless, living with my plastic bags like a hobo, begging for money. I'm extremely nervous inside the house, so worried I'll be noticed by a neighbor. I feel physically sick and nearly paralyzed. But I work fast. When I'm done, I go back to school, and there, I recruit a few willing friends to help me with the garbage bags. My friends come to my aid.

We drag the black bags to Elisabeth's house, where we make a preliminary plan. We call all the upperclassman boys we know, the ones who have cars, and ask them to drive me to the bus station. They're all sympathetic to my situation, but none agree.

As night falls, Elisabeth's mother gets wise to our plan, and Elisabeth pleads with her to let me stay. From her bedroom, our friend Whitney and I can hear the two of them screaming at each other. Elisabeth's mother cries, compassionate but powerless, "I'm sorry! I'm sorry! We can't help her!"

I know they both want to help, and my heart breaks for them, especially for Elisabeth, who's fighting for me. I am causing them to fight and scream and cry, and I feel so guilty.

Whitney suggests we call her parents and ask them what to do. When she hangs up, she smiles and nods reassuringly. After a short while, Whitney's parents arrive at Elisabeth's house. They pick up Whitney and me, loading my garbage bags into their van. I'll spend at least the night with them while I figure out my next move. Maybe I won't have to live on the street since Whitney's parents are helping; I'm so grateful for this reprieve.

The next day, I go to school as usual. On my way to first period, I stop at the front office to let the secretary know that I'm running away from home and that I'm staying at Whitney's. I've heard you could be sent to juvenile hall for running away, so I want to be as transparent as possible. I want the principal to understand that I'm not a criminal—I'm only staying away from home while my mother's away. And, of course, I'm still planning to attend school every day. I lay all the facts out plainly to the secretary, then head off to class. I don't want any trouble with the law.

A couple of hours later, I'm called out of class. The principal wants to see me. I'm led into a back room where there's a whole delegation waiting for me: teachers, counselors, coaches, the vice principal, and the plainclothes cops who patrol the smoking section of our school. The kids call these officers narcs. The main narc looks at me with sad eyes, which is weird given how he usually strides around school like a badass, scruffy-bearded stoner in a flannel shirt. In this room, he looks like the most empathetic adult, so sad for me. It breaks my heart. This is all so strange!

The ten or so adults are seated in folding chairs in a semicircle. They wear solemn faces, and most avoid making eye contact with me. I've never really seen adults look like this before. Aren't they all sort of overreacting? I mean, this is serious, but why is everyone acting so sad and mournful? The girls' basketball coach, Mr. Johnson, is here too, looking sheepish and forlorn. It's almost funny to see these tough men looking at me with such dismal kindness. I realize I've never seen emotions like these in a man's eyes before. I'm thrown way off by it, unfamiliar as it is. My guard is immediately up.

The vice principal asks me questions that are both cautious and vague. It's like she's trying not to show emotion or put words in my mouth, which is a shame because I do not have any words of my own. This is much more than I bargained for, and I start to drift away, tuning out the many voices around me. Out of all these adults, not one can guide me through this. I feel alone and in way over my head. The tension is too great. I'm still, after all, only a child.

All these eyes on me have me dizzy. Did they explain what's going on? I can't hear anything over the deafening buzz in my ears. I offer perfunctory answers, all truthful but with none of the elaboration they obviously want. I didn't turn myself in as a runaway expecting an inquisition. I have not prepared for this, and the stress is overwhelming, so I shut down.

My eyes wander, and I catch glimpses of my classmates passing through the hall outside. It must already be time for the break between classes. *Can they see me through the glass*, I wonder, *or is it a mirror on the other side?* I worry they can see my frightened face. I'm not crying, but it's taking all my energy not to. The buzzing in my ears gets louder and louder.

And then, suddenly, I'm released. The day is somehow over. I go back to Whitney's house and relax there. Her family is fun and kooky and kind. They have an eclectic house, and the kids—Whitney, Brent, and Brady— seem happy and loved. I want to stay. I feel like a normal kid in their house.

When her parents go out, Whitney and I turn up some music, lock her brothers outside, and call boys from school to flirt with them and find out if they like us. I had no idea being a teenager could be so much *fun!* I wonder if this could be a solution in itself. *Can I just stay here and live with them?*

Two weeks go by, and my mother returns from France. Whitney's mother arranges for counselors to meet with us and help us talk things out, which happens in Whitney's living room.

I already told Whitney and her mom that Sal hits me almost every day. They both listened carefully, and when I offered a few examples of the abuse, Whitney's mom stopped me.

"I don't need to hear any more," she said, stroking my back.

It was the first time an adult took my story seriously.

When my mother arrives for the meeting with the counselors, she is livid. She storms into the house and, with tight lips, orders me to get my things and meet her in the car. I look at the counselors seated politely in the living room, desperate for them to intervene.

Then my mother lashes out at Whitney's mother. "You have *no right* to get involved in other people's business! Who do you think you are?" she shouts.

She announces that she won't speak to any counselor.

"Susan, I said *get your things!*"

I look at Whitney's mom and at the counselors, but they just look back at me. I know by the looks on their faces that there's nothing they can do. My mother is going to explode if I don't go with her. I go upstairs, get my things, and follow her out to the car.

When we get to the house, I'm sent to my room to wait until Sal gets home. This will all be discussed at the dinner table, the official place for announcements and threats of forthcoming punishments.

In a while, I'm called down for dinner. It's sunny outside. My brother Evan sets the table, and Sal takes his seat at its head. I take mine to his left. The table is round and made of glass. My little brother Giovanni is about six years old. My baby brother Matteo is four and is sitting in his booster seat. Evan sits across from me. As I watch Evan take on some of my chores around the dinner table, I realize they've gotten along fine without me.

Sal begins to serve himself. As he carves into his steak, the family is quiet. He finally looks up at me, his knife in his hand.

"I bet you think you are so smart," he says.

There's a long pause as he chews his meat, a smirk on his face and his eyes glinting with satisfaction. He stares at me as he chews.

"You wanted to call the cops on me? You wanted to turn me in? Do you know who I am?"

I look at my plate, push some lettuce around. I am so ashamed to be home, a failure.

"I know all the cops in this area. I know the chief of police. Do you think the cops are going to listen to *you?*"

I stare back, mortified.

"Well," he says, his smile growing, "Child Protective Services came to my office. Do you think they believed *you?* CPS came to the house. They talked to your brother, and do you know what he said? Your brother told them you are a liar. HA!"

I'm frozen in terror.

"Did you hear me? Now *everyone knows you are a liar!*"

He gives me time to let this all sink in. He speaks slowly and clearly, using simple sentences so that I will understand every word.

"Your teachers know, your coaches know, the parents know, *everyone knows!*" Sal laughs boisterously in his victory.

My eyes settle back on the food in front of me as I absorb this information. Yes, he is correct. Everyone thinks I am a liar now.

I am devastated but not destroyed. I understand what has happened.

I understand that I have been proven a liar, both by my brother's assertion that we've never been hit and by Sal's friendship with local police. I understand my friends' parents now think I lied. Everyone thinks I lied—except the people around this very table. My mother, my brothers, Sal—they know I am not a liar.

There's a particular loneliness that comes when you're carrying the truth and no one else seems willing to hold it with you. I'm not just isolated physically—I'm standing on my own moral island, watching the people around me choose comfort, neutrality, self-preservation, or politeness over what is right. It makes me doubt myself more than I care to admit. But in the end, I know what I felt. I know what I lived. And that knowing becomes its own anchor.

I know that it will be even harder to get away now because I played my hand and lost. I still have two more years of high school ahead of me. That's a long time. Unimaginable, really.

I have no words for Sal. I am excused from the table, sent to my room, and told to think about what I have done. I go, solemnly. And I wait. I slip headphones over my ears and press play on my Sony Walkman, loaded with the cassette tape I made by pirating all my favorite songs from the radio. I keep myself busy by drawing boxes with colored markers, making my own calendar on a pad of plain white paper so I can count the months until my eighteenth birthday. And I wait.

While I draw, I wonder why my brother lied to the police. Maybe he lied because I never came to save him when I heard his screams. When I heard him crying out, "Help me!"

When the beatings happened, I didn't cry out because I knew no one was listening. My mother sure wasn't.

But when it was my brother's turn, he knew I was there. He knew I was listening.

Night after night, my brother cried out for me. And I did nothing but lie in my bed and try not to breathe.

CHAPTER FOURTEEN

HERE FOR A REASON

"There is one simple wrong with you—you think you have plenty of time."

Carlos Castaneda

1997—SAN JOSE, CALIFORNIA

When I first seek a lawyer to represent me in the custody case, I find a kind female family law attorney I like. She reads Cliff's initial petition and says, "These are heavy accusations. How do you want to handle this?"

I tell her, "I just want it to be handled quickly. I'm sure the court will see through these false accusations. There's nothing there."

She looks at me with concern. "You might be in for a fight. They're coming on strong. We could counter. Tell the court about his DUI, gather evidence of how he treated you while you were married?"

"No," I say. "I don't want to fight. Let's settle this calmly. I want to move on."

I still assumed lawyers were like Link, my lawyer from the divorce—effortless and winning. It hadn't occurred to me yet that Link was an outlier: a well-respected criminal defense and civil litigator who probably

took my case as a favor to my stepfather. He handled it with a kind of protective ease, and I thought that was simply how the legal system worked. I
was about to find out that family court could be a different world entirely.

Cliff's custody suit reframes reality. Suddenly *he* is the one needing
protection—from an accusation I never made. I took Emma to a county
therapist because she was showing signs of stress. That was the only responsible thing I could do. I never accused him of abuse. But in the cli-
mate of the 1990s, a mother raising even the faintest concern through
proper channels could be painted as unstable or vindictive. The legal and
cultural context at the time tended to pathologize women who pushed
for protective action, especially when they took allegations to court or
therapists. His attorney builds their case around disproving something I
never said and using it as proof that I'm unfit. This case is a distraction.

At night, when fear creeps in, I repeat to myself: There's no evidence
against me. I'm a devoted mother. Cliff has a past restraining order on
file and a DUI. I've never broken the law—except that one time I stole
a Snickers bar from 7/11. I got caught and learned from it, and I was
twelve. Other than that, what evidence did they have to prove I was even
slightly unfit? The fact that I was working outside the home was one of
the true statements supporting their case, but that's not illegal. Surely
the court will see through this.

What they are suggesting—in language carefully cloaked in legalese
—is that I am coaching Emma to make abuse claims because of my own
unresolved trauma. It's a brutal sleight of hand. I had lived through violence, but not sexual abuse, and never made such a claim against Cliff.
Emma was asking for help. I answered. That's it. But the accusation they
fear hasn't been spoken aloud—and they're preparing their defense anyway. It feels like the move of a guilty man. But court is still ahead. We
are in mediation.

I am sprinting across cracking ice—motherhood on one side, ambition on the other, failure at both lying below. I don't have time to process
how I feel. I just keep moving.

1997—SAN FRANCISCO, CALIFORNIA

Meanwhile, Webgrrls continues to thrive. We all know the cliché moment when the main character spontaneously jumps up on a table and gives an impassioned, unrehearsed speech, which then starts a movement. I never imagined myself as that person, but one night, at the Dark Horse Café on Pacific Avenue before dozens of women, that's exactly what I do.

That foggy San Francisco evening, before my coffee table speech, young women line up in queues like military ants for espresso. The artfully dim, almost Victorian-styled room is bursting with chatter, activity, and excitement. Overstuffed furniture and jewel-toned velvet chaise lounges adorn the narrow café.

I do not intend to speak. However, our plans for a quiet coffee klatch to get to know some of our new members are soon overruled by the buzzing white noise of the Italian milk steamer, which makes conversation unintelligible. Looking through the crowd, I feel this event needs some soul. Some heart. I spontaneously climb up over a classic leather club chair onto a coffee table to rally the attendees and to reward their expectations. In short, I want to give these early "grrls" of the dot-com nineties what they came for: inspiration and community.

I whistle once—inelegantly with my fingers—over the cacophony, and eyeballs peek up at me over rustic vessels of espresso. The room falls silent. I clasp my hands in front of me and attempt to earn their attention.

First, I welcome everyone and acknowledge the unexpected size of the crowd. Standing on the coffee table, I feel awash in discomfort at towering over everyone, but I go on, explaining the mission of the San Francisco Webgrrls chapter and the community we are building together, our answer to the old boys' network.

I make a light joke about drinking espresso at night—"Our morning," I say to laughter, "so we can stay up and bang on our keyboards 'til dawn." Then I thank the owner of the Dark Horse, a young, clean-cut guy with dark hair and sweet boyish cheeks.

The women applaud our host.

As silence resumes, faces both knowing and seeking return their attention to me, many through retro, pointy eyeglass frames. I sense a rare but welcome companion enter the room: the spirit that says history is happening right now—that spirit boosts me and gives me words.

"In most cases, each of us is the only woman in the room—where we work, where we pitch, where we write code. But in this room, and on the Webgrrls list, we are not alone. We are joined by our sisters, women who walk the same path, face the same challenges, and share the same dreams. We are here to lift each other up, to lend our strength when another falters."

This type of affirmational, sentimental reverie, which today seems like cringeworthy kumbaya, was necessary. At the time, being the only woman on the coding team often didn't feel so much groundbreaking as it felt like a throwback. Would we be admitted to the boys' club? The answer was no, not really. The tech of the day was cutting edge, but the culture was stuck in the Stone Age. If you wore lipstick, someone would comment. If you didn't, someone would notice. You had to be fluent in a language of exclusion and still manage to speak with authority. The office was part locker room, part control room. While the boys went off after work to collaborate over Jolt Colas and talk about *Babylon 5*, we walked home through the alleys of San Francisco alone. But in that room, we were no longer walking alone. There was no built-in network for us, so we were creating it. Together.

"And here, in this community, if one of us has a code crash at 3:00 a.m., another will check her email, roll up her sleeves, and help her through it."

I pause, not knowing what I will say next but very much feeling the spirit, and I look directly into the eyes of as many of the women as I can—to the center of the crowd, to the shy ones standing against the back wall. I hold their gazes. I know that I am a technology pioneer, but so are they. I am not alone.

"We are not just workers, coders, or creators—we are pioneers. Every single one of us. Simply by being here, by daring to show up in spaces where we're often the only ones with ovaries, we are breaking ground, laying the foundation for others to follow. I want you to take a moment, right now, and truly acknowledge that. You are not just a part of the future. You *are* the future. Together, we're building something no one can ignore."

I said what needed saying, and this was the message that hit the mark. Bullseye.

I invite the women to applaud themselves, and they do. Some shout, "Let's do this!"

I climb down from the table. Throngs of brilliant, educated, enthusiastic young women surround me, all handing me their business cards and asking how they can volunteer. I spend the rest of the night telling them to join the email list and attend the monthly events.

"Come with questions and requests," I advise them. "Our speakers are the women who've come before us, and they're often still at the top of their game—with hands open to grab ours or to hold open a door if they can."

The young women nod, their eyes glued to me, soaking up my words.

"Tell us what you need," I continue. "A contact at Kleiner Perkins for your start-up? We'll help. And tell us what you have to give. Trying to fill a job in the Ruby coding division, but you've got only male applicants so far? Put the job ad on the list. You'll find qualified coders here. That's the credo of Webgrrls. That's what we do."

On my daily commute to work, I step off the catamaran on the Tiburon-San Francisco ferry route. I slip my subway-folded copy of the *San Francisco Chronicle*'s business section into my handbag next to my laptop and sling the leather bag over my shoulder for the brisk walk from the historic Ferry Building through Embarcadero Plaza to Drumm, up Jackson Street and west to the Wordcasters office at 400 Pacific Avenue in Jackson Square, just a two-minute walk from the landmark Transamerica Pyramid. I stop for a latte at full power and full fat to give me the superhuman strength I need in this early dot-com boom.

Inside our historic brick building, I bound up the stairs and open the blinds before starting up my desktop computer. Between sips of my supercaffeinated creamy beverage and bites of the delicious Godiva chocolate truffles I keep at my desk, I log on. I power through emails and phone calls until finally looking up from my screen around 1:00 p.m. The individually wrapped sticks of string cheese I keep in our lunchroom fridge for the kids are my go-to snack, wrapped in a slice of deli ham or turkey. I climb out of the window to the fire escape and take a break to eat while gazing down at the street below.

Some female venture capitalists, like Ann Winblad of Hummer Winblad, software investor and well-known ex-girlfriend of Bill Gates, and Darlene Mann, an advocate for increasing female representation in tech leadership and general partner at ONSET Ventures, have started to notice me. Both Winblad and Mann are influential figures in the venture capital landscape of the late 1990s, contributing to the growth and success of numerous technology start-ups during a pivotal era in the industry. I am invited to pitch to a handful of mainstream venture capitalist firms in San Francisco and Menlo Park—especially on Sand Hill Road, for those on the critical artery of tech investment.

My bio in the business plan for these pitch meetings doesn't list my education, but by now, I'm prepared for those questions.

"Where did you go to school?" asks the partner rifling through papers at the conference table, not hiding his annoyance.

"I didn't go to college," I say with a practiced, carefree mix of mystery, pride, nonchalance, and a touch of defiance, daring him to ask the next question. I twirl my pen and recline in the conference chair, holding eye contact.

"You ... didn't go to business school?" he asks. "You don't have a degree in ... anything?"

"I launched companies instead. If I had gone to university, I would just be starting out now."

I admire and respect those who've been accepted to university, paid or borrowed a fortune for college tuition, and completed an undeniably rigorous educational process that instilled in them a knowledge of their field and, perhaps more importantly, the connections to alumni to help them climb the ranks of business post-grad. But their education only covers what's been done before. While they were learning about history, I was making it.

The investors and reporters know a company needs at least one entrepreneur like me. One who sees things others don't, who envisions building things other people don't think are possible. Still, not everyone feels that way. Back at my mom's house, when I share my wins with them around the dinner table, I am misunderstood. They say I always have an answer for everything. I am delusional—Sal calls it "blind optimism." They say that I manipulate scenarios to my benefit. These words sting, but for so long, I just keep on trying to prove myself to them, still wanting the love and approval they never gave me.

If I can't win them over, I can at least do everything in my power to improve myself and my business. I have huge weaknesses—sales and financials. Most venture capitalists don't understand anything about the internet yet. They are business dudes my parents' age, not tech geeks. They know so little that they don't even notice when a slam-dunk concept, a proven business model, and groundbreaking technology are smacking them in the face. Many VCs are just copying what other VCs are doing, hoping to catch a ride on a flying carpet they don't understand or know how to steer.

Soon, I am asking for $20 million in Series A financing, but I'm taken aback when a seasoned venture capitalist tells me his firm is declining to invest in Wordcasters because of a conflict with one of their other companies, which offers a similar service. I ask who he's talking about, and his answer is Pointcast.

I can feel my blood boil in my veins.

"We are not a screensaver pushing out newspaper headlines!" I gasp.

He stares back at me blankly, embarrassed by my visible incredulity at his lack of tech knowledge. He is annoyed, and I think I'm burning a bridge.

"Besides, Wordcasters is profitable! We have revenue. Pointcast doesn't."

"True," he says. "But we're investing in the new economy."

This is a nice way of saying they want to be in the game so badly that they'll jump on trends regardless of their proof of market.

At this time, start-ups inflate their financials with insane phantom revenue—projected sales numbers that have not been collected on—and the VCs don't care. They'll throw $50 million at a start-up, buy users, blast their marketing and court the media for exposure, then push an initial public offering within eighteen months and make their money back through the IPO. Actual revenue and profit don't matter.

Clearly, I don't speak the language of financial projections, even though everyone knows they're 100 percent bullshit. Everyone is making up numbers for products and companies based on technology that has yet to be invented and imagining markets that are yet to exist. The whole song and dance is so stupid and pointless. How big will the market be for internet communication in twenty years? I don't know—bigger than the biggest number a five-year-old can make up? Like bajillion gazillion? So many digits, you can't fit them onto your silly Excel spreadsheet. I know this "new economy" is a bust. How can these expert financiers tell me that the traditional economic rules no longer apply? I'm not a finance whiz, but this garbage thinking is obviously not grounded in reality. I am incensed.

It's not that I don't understand the *real* financials. I know how to price my product, manage all the revenue streams, and calculate the money I'm saving my clients compared to the alternatives they used before. I know how to adjust the expenses for the marketing and sales teams, equipment infrastructure, overhead, taxes, insurance, and outside consultants—for each product sector at any given time. I know exactly how much money we have in the bank and how much in accounts receivable because I manage the books myself.

Like me, my company does not play by conventional rules, but we are both the better for it. We are profitable, so I can continue to bootstrap and retain majority ownership a little longer and wait for the major players to catch up.

CHAPTER FIFTEEN

A SEAT AT THE TABLE

"The value of an idea lies in the using of it."

Thomas Edison

1997—BOSTON, MASSACHUSETTS

I scan the room for an empty seat at the Boston Convention and Exhibition Center after filling my plate at the well-appointed buffet. Among the sea of men in casual business attire or pullover logo fleeces at the 1997 Voice on the Net (VON) tech conference, there is only one open spot remaining, and it's directly in front of me at a round table for ten. At twenty-eight years old, I look easily ten years younger than I am and am keenly aware of how out of place I appear to be. I wonder if my dress is wrinkled and if I'm sweating through the fabric. It isn't, and I'm not.

I recognize the white-bearded man in the three-piece suit inviting me with one elegant hand to join the table. It's one of the fathers of the internet, Vint Cerf, a modern-day inventor—Benjamin Franklin to geeks like me. Cerf is the codeveloper of the ARPANET and the DARPA and the author of the TCP/IP communication protocols that enable the internet itself to exist. He's the architect of the first commercial email system for

MCI, back in the 1980s, and is currently working on the Interplanetary Internet at NASA's Jet Propulsion Laboratory.

I'm naturally intimidated, but I can't humbly shrink away now that he has me in an eye lock. I take the seat across from him, worried that I'm entirely out of place at his table.

"Join us. Don't be shy," Vint says, beckoning me formally but warmly. Then he turns to the service staff member at his shoulder, who is waiting for his beverage order. Vint asks for hot water and a teacup. "Not a cup of tea—just a pot of very hot water, a teacup, and two lemon wedges, please."

Then he turns back to us at the table.

"Best thing you can do for your health," he announces to our nine puzzled faces. "Especially first thing in the morning!"

I watch in awe as his attention turns to me. His eyes pierce right through me.

"And you are?" he says.

"I'm Susan Quinn."

I feel my face get hot.

"And what do you do, Susan Quinn?"

"My company, Wordcasters, is here streaming the conference live on the net. We just live-streamed your presentation, actually."

Cerf seems surprised. "Tell me more. You live streamed . . . text? Of my presentation . . . just now? I'm intrigued."

Vint sets down his cup of hot water and folds his hands, leaning in. Everyone at the table looks at me. His interest is real, which puts me at ease.

"I invented TextCast, and Wordcasters livestreams voice to text in real-time over the internet at three hundred words per minute with 99 percent accuracy and three seconds latency."

I explain that we're streaming the sold-out conference so people who couldn't get tickets can attend live online.

I discuss how the text is delivered and explain the procedure for the end user to log in and follow along with the conference. Vint asks smart questions to understand my process, and when he does, he immediately starts on practical applications.

"This has incredible use cases for information accessibility globally and remarkable potential for individuals with hearing impairments," he says.

Vint's wife is deaf, and he has significant hearing loss as well, he explains, motioning to his hearing aid, which I hadn't noticed.

I'm thrilled that he cares about access, too, and not just profitability. Wordcasters enables access to information for anyone regardless of their physical location, bandwidth, or sensory ability.

"You see," I continue, "the text stream is delivered in such a small packet that anyone in the world with an internet connection and a computer can receive it in real time, no matter their internet speed. We put up an archive transcript within the hour, which can be translated into any language and is immediately searchable across search engines, unlike audio tapes, which take weeks to prepare and distribute—not to mention that they're expensive to replicate and mail out. By the time anyone gets the tapes, the information is already old."

Vint smiles. "I would love to discuss the applications in more detail," he says. "Please call my office, and we'll set something up with my group at MCI."

He hands me his business card and takes mine.

I completely forgot to eat, and now I have to get back to the main stage for the next keynote before everyone else does. My team producing the livestream today is in Dallas, Texas. My job on-site is ensuring my team has access to the audio feed via my special transmission box, which operates over a dedicated phone line I set up with the A/V techs at the conference center. I have to fly out with it myself because I only have one device, and it's too precious to let out of my custody. It's an old, discontinued Gentner box that I picked up from RadioShack, so I have to fly to every event and hand-deliver it until I can figure out how to make more of them. I bid farewell for the moment to everyone at my table and dash out so that I can go turn it on and start the feed before the next keynote.

I bounce back to the main stage, feeling elated. It turns out I did belong at the table.

Everyone knows Vint is a busy man, and MCI is a telecom company that doesn't need my technology. At best, MCI could fund my development as an experimental investment in the future. I'd be thrilled to continue the conversation with Vint with or without MCI, though. In my follow-up email, I invite him to sit on my advisory board, and I ask for his insights on evolving technology and partners he could connect me to. I admire Vint and have ideas to make him the first avatar livestreamed alongside text, sort of like an AI hologram.

I'm flying high as I wrap up the conference and head home. The professional advances I've made over the last few years continue to surprise and impress me. Every time I feel I've reached a new zenith, I somehow reach another one.

CHAPTER SIXTEEN

BILLION-DOLLAR BLUFF

"With a little bit of nothing, you can make the possible happen."

Oz

I am standing nervously in the hallway of Hambrecht & Quist's PlaNET.wall.street investment event at the top of Nob Hill. I used to dress for the part—blazers, old lady suits, a short bob haircut I thought might help me look older—to blend into boardrooms and tech panels. But it always felt like I was dressed in an awkward costume. Over time, I stopped wearing pantyhose and old lady suits and ended my efforts to look like someone older. My hair grew past my shoulders. I bought A-line dresses in light blues and solid black and wore sleeveless fitted tops with wide-leg slacks from Banana Republic. No more boxy lines, shoulder pads, or stiff collars. Easy, flowing, clean silhouettes and fabrics that let me breathe. Despite my nerves, I finally feel like I belong.

Everyone is here. In the halls of the Ritz-Carlton Hotel in San Francisco, I recognize Kim Polese, CEO of Marimba and current darling of the season. She was a speaker at one of our Webgrrls events, and she personally helped me find my programmer, Patrick Chan, when I called her up to ask her who would be the best at improving my software using the Java programming language. Kim is surrounded like the popular cheerleader, so I don't go over to say "hi" like I should. Eric Schmidt of Sun Microsystems walks by and glances at my monitor showing the live text of what is happening inside the conference room steps away. Eric will later become the celebrated CEO of something no one has heard of yet, Google, but at this point in time, Sergey and Larry are still sitting in their Stanford dorm room coming up with the brainchild they first called "Backrub."

According to internet history, in 1996, Larry Page and Sergey Brin, graduate students at Stanford University, developed a search engine called "Backrub." It analyzed backlinks to determine the importance of web pages —a novel approach at the time. In 1997, they decided the name "Backrub" wasn't quite right. They brainstormed alternatives and landed on Google, a play on the word "googol" (which means 10^{100}—a one followed by one hundred zeros), reflecting their mission to organize a massive amount of information. I recall the number googol as described by Carl Sagan on PBS in the early 1980s when my little brother Evan was obsessed with trapezoids and large numbers like googolplex and infinity.

Venture capitalists, the world's leading institutional investors, angels, and the top forty CEOs in emerging internet technology—they are all there. The cream of the crop.

Mark Cuban, who I know from the Usenet groups we were both previously active in, is just as much a blowhard instigator and flame war starter in person as he was in those early chat rooms. Mark is talking to everyone in the hallways of the breakout sessions—and this is before anyone knows him. He is talking. I am *doing*.

It's now time for the last, and the most prestigious, panel discussion of the conference. I'm hoping to prove the viability of the interactivity features of my platform right here in front of everyone who matters in internet technology.

Just then, I hear a sonorous pinging sound—a question! A real live question from the online audience! It works. My hand shaking, I scribble the question on a piece of scrap paper and run into the conference room. I have to push through human obstacles just to get through the door since it's standing room only, and I can barely squeeze in among the business suit-clad attendees towering above me in my baby blue A-line dress and nude ballet flats. The room is dark with spotlights on the stage where the panel is wrapping up the Q&A, and my eyes struggle to adjust.

I hear the moderator say, "We'll take one more question . . ."

As he scans the room, I know, I just *know* as my hand shoots up high from the back that he is picking me. It's my time, and I know it. And he does too.

After he points at me, one of the runners hands me a microphone.

"I'm Susan Quinn," I say, manufacturing infinite confidence so I'm unmistakable, "and I'm here with a question from the live audience online watching on Wordcasters.com."

I can see the panelists, Marc Andreessen, Eric Schmidt, Halsey Minor, and Ed Kozel, all look up when they hear what I am saying. They scrunch their faces against the lights to see who I am. Marc Andreessen holds his hand up to shade his eyes from the bright lights, squinting to see me. I take a moment to let them see my face. I then look at the scrap of paper in my shaking hand. This is my moment. I speak clearly and loudly into the microphone.

"This question is from Allan Price, who is watching the discussion live from Tandem Computers HQ in Cupertino, California . . ."

I read from the scribbled note I just captured from my bulky ecru-colored computer monitor in the hallway, the one I carried here at 6 a.m. and set up on a banquet table next to the candy bar table.

As I read the question, I can sense people wondering, *What does she mean "watching live from Cupertino?"* This has never happened before.

When the spotlight finally swung my way, I thought I'd feel whole. But I don't. The recognition feels strange, incomplete. They see the success, the invention, the branding—but not the mother with a court summons under her laptop or the woman pushing down tears on the ferry commute.

They notice.

But then they forget.

S oon after the event, I receive a memorable contact from Patrick Sea-
man. He introduces himself as the chief technology officer of Audionet, a company I'm familiar with—launched by Todd Wagner
and Mark Cuban.

Oh, boy, here we go. Mark Cuban has been circling me and my company since 1996. At that time, he was just a muscle-bound, fast-talking
guy in Dallas trying to make noise in the new "streaming media" world,
which barely existed yet. In my opinion, Mark was more hot air than innovation—a self-made carnival barker in the Wild West of the web.

Audionet is Cuban's company. It claims to broadcast live audio streams
of sports games online—just the audio from local radio stations, rerouted
through the internet. To me, this isn't real technology. It's rebroadcasting
someone else's signal with a digital extension cord. Cuban didn't invent a
platform. He didn't solve a problem. He just yells louder than anyone else.

Meanwhile, Silicon Valley is just waking up to the possibility of the
internet as a media channel. Netscape went public a year ago, sparking a
gold rush. Everyone wants in, and no one knows exactly how it will work.
Streaming media is still clunky and unreliable. RealAudio has barely gotten off the ground. RealVideo is a rumor. Even Microsoft is struggling
to figure it out.

Mark Cuban isn't a tech guy, but he draws a lot of attention to himself
—fighting with RealNetworks over exclusivity deals, fighting with radio
stations about copyrights, fighting with ad networks over banner rights.
In these days, noise and bravado are currency, and Cuban knows how to
spend it.

Patrick is polite, but he doesn't strike me as a real technologist. He
seems more like a stand-in CTO—someone with a thin layer of credibility to fill the position at Audionet while Mark and his partner, Todd
Wagner, work the business angles.

The company, Patrick explains, has a proposal for me.

"See, Mark and Todd want to buy your company, Wordcasters. Would
you like to discuss the price?"

Uh, sure.

I figure anything is for sale at the right price. I have investors to answer to, after all. I love what I'm building, but I know my job is to grow it, succeed, and exit well. Wordcasters is profitable, scaling in the corporate sector, and has real paying clients—unlike most streaming start-ups, including Audionet.

During our initial talks, Patrick and Todd explain that my scrappy little company—a *real-time text-streaming platform*—will become a key part of Audionet's strategy to show future investors that they can generate actual revenue. Corporate revenue. Microsoft pays me $10,000 per hour with a minimum of three hours for corporate events. Wordcasters is solving a real problem: how to bring live events to remote audiences *reliably*—not just shouting about it.

It becomes obvious to me that they need Wordcasters—badly. Without a real product and without a revenue stream, they are just another wannabe start-up in a sea of wannabe start-ups.

Then comes the offer.

They want to acquire Wordcasters in exchange for stock—10 percent of the company.

Stock, not cash? Stock in a company that hasn't invented anything, has no meaningful revenue, and is essentially riding a thin legal line by rebroadcasting sports radio over the internet?

Oh, hell no.

I briefly contemplate what percentage of their stock I should ask for . . . 20? 40? Nah, they have nothing and are going nowhere. I decline the offer outright.

"No, thank you. Goodbye."

Two years later, I'm drinking my morning coffee when I see the headline: "YAHOO! BUYS BROADCAST.COM FOR $5.7 BILLION."

In those two years that felt like a split second, Mark Cuban rebranded Audionet as Broadcast.com, then rode the dot-com wave all the way to a nearly $6 billion cash-out.

I do the math. What's 10 percent of $6 billion?

$600 million.

Six. Hundred. Million.

Oops.

But was I wrong, really?

Broadcast.com is never heard from again. Yahoo! absorbs it, struggles to justify the acquisition, and eventually buries it.

And Mark Cuban? He buys a private jet, parties, buys the Dallas Mavericks, and exits tech altogether, for a long while, at least.

He didn't build the future. He just knew how to cash out before anyone else realized he was bluffing. If nothing else, he paved the way for many of the snake oil salesmen turned tech hype guys we'd see decades later. Like Elon Musk—or whoever else is shouting the loudest in hopes that the money will follow.

So did I make a mistake?

Maybe.

But I still believe in what I'm building. Not just the noise. Not just the moment. The future.

Soon thereafter, the dot-com bubble will burst—thanks to companies that do nothing and a "new economy" that relies on IPOs as exit strategies. Companies that don't need revenue? Yeah, that never made sense to me.

Sure, I could be swimming in millions by age thirty and moving on to my next invention. But in the end, I'm right about the new economy. It isn't sustainable. I want to be a part of something that is.

CHAPTER SEVENTEEN

SUMMIT AND ABYSS

"To keep our faces toward change, and behave like free spirits
in the presence of fate, is strength undefeatable."

Helen Keller

1997—SEATTLE, WASHINGTON

O nce I've generated several live transcripts for Bill Gates's keynote speeches, I'm summoned to appear at the Microsoft Headquarters in Seattle. It's a year after Gates first became aware of me at Microsoft's Professional Developers Conference.

The invitation is technically phrased as a request, but I know better.

With my other high-level connections—like Larry Irving, President Clinton's top tech advisor—I know the offer to connect in the future is unlikely to become a reality. But it's different with Microsoft. If they want to know what you're doing, they find out what you're doing.

The general consensus among tech start-ups at the precipice of success is that Microsoft will pull apart your code, devour your intellectual property, and either force you to sell your company to them or crush you into oblivion, leaving no trace of your existence. Thus, getting the call from Microsoft is as exciting as it is potentially devastating. It means I'm doing something worth stealing.

I'm afraid of revealing too much of my intellectual property and allowing Microsoft to steal my work without even having to buy my company. They do not sign non-disclosure agreements, just like all the venture capitalists, so I'm used to it, but the fears still nag at me. I worry that I'll freeze during the presentation, sweat through my clothes, or have a stroke and die. I don't doubt my technology, but I am still concerned about my nerves getting to me. And what if they tell me about a competitor who has more industry connections and financial backing? Surely they wouldn't want to talk to me if that were the case.

Even in the best-case scenario, if Microsoft loves the technology and asks me to move to Seattle and develop the tech for them, I'm not sure that's what I want. I'd have to give up my life and my business to work for someone else. I'd be trapped in a cubicle and chained to my desk with the proverbial golden handcuffs.

In the end, my overwhelming faith and enthusiasm for this work win out. I know this technology is not just legitimate but groundbreaking, and I'll get to tell the biggest players in the industry just that.

Leaving the girls at home with Jake, I fly up to Seattle the night before the meeting and check into my hotel after dark. I'm nervous because I don't know what to expect from the presentation. Normally, I know exactly what and why I'm pitching and what my desired outcome for any meeting is. But in this situation, I have no idea what success looks like or what exactly Microsoft has in mind.

I have ten printed pitch decks to pass out at the meeting. Hopefully, they'll keep people from looking directly at me the whole time and will buy me some time if I get stuck. Going through the deck in my hotel room, however, I find a typo, so I look up twenty-four-hour Kinko's print-shops in downtown Seattle so I can reprint the incorrect page. The only available Kinko's won't open until 6:00 the next morning, so I go to sleep. Except, of course, I don't sleep at all. I toss and turn, going over my presentation in my head, then head to the store in the pitch-black early morning.

When the sleepy Kinko's employee finally unlocks the door, my face and hands feel frozen, and my brain is mush from lack of sleep. I print the pages, but they don't match the colors of the existing pitch deck. I sigh. This whole errand has been pointless from the beginning, but I need to channel my anxiety into something approximating productivity. I race back to the hotel to get ready, then meet Victor Anderson, a new hire I've brought along for his expertise and experience, in the Microsoft lobby.

Victor is calm and collected while I'm buzzing from several Starbucks lattes. Finally, I start to feel my physical exhaustion wear off a bit. Victor has been invaluable in explaining to potential partners that Wordcasters has the skill, experience, and industry knowledge necessary to succeed. But the pitching will be up to me. I cannot train Victor to answer questions with the spark, deep knowledge, and confidence I have.

During the presentation, not a single person looks at the printed materials I agonized over. Instead, they stare at me with eager and enthusiastic eyes. The group of twenty or so people is friendly and interested—and quite a bit younger and savvier than any of the VC panels I've pitched to in Silicon Valley. I deftly answer every one of their questions, filing away my fear as I do. I have to remember this feeling so that next time, I don't need to worry as much. I've got what it takes, and I don't need to waste a second more doubting my abilities when I could be working on new ideas and products instead.

After the first presentation, Victor and I wait for an hour in an empty conference room for two of Bill Gates's top advisors, who request a more in-depth discussion. My nervousness returns because I still don't know what they want. I don't know what *I* want. Why are we even here?

In the end, nothing ever comes from this meeting. Our tech is credible enough to attract Microsoft's interest while simultaneously allowing us to avoid any semi-hostile takeovers. The pitch process also perfectly illustrates the industry at this time. Everyone is on the hunt for the next hot thing but keeping their own cards close to the chest. There's lots of excitement but not always the infrastructure, knowledge, or resources to follow through and turn big ideas into reality.

All in all, it's a thrilling experience and another powerful lesson, illustrating my ability to stand before the best in the industry. I'll get better at it the more I practice.

Waiting for me at home, though, is the comedown. It looks innocuous at first in its humdrum FedEx envelope, which Jake dutifully signed for in my absence. Inside is a subpoena to appear in San Jose family court on a contempt of court violation. Apparently, I missed a hearing I didn't know about. Cliff's lawyer says I was served a notice by fax, but I never received any fax. Now I need to appear before the judge, and he's not happy.

Being taken to court for contempt—for trying to protect my daughter —is one of the most disorienting experiences of my life. I'm being called in to face the judge to defend myself for not following procedure, but it's a trap. The hardest part is that no one in that courtroom seems to care what is true. It's about appearances, compliance, optics. I have to sit there, heart pounding, while people who don't know me decide whether I'm "stable" enough to keep my own child. I'm not just fighting for custody— I'm fighting for my reality to be recognized.

When it comes to these court proceedings, this feeling of being untethered from reality is becoming horribly familiar. I need to be strong, but I begin to allow myself to consider the doubt that enters my thoughts frequently. Do I truly have the capacity to succeed at both business and motherhood? I know that thought I did when I started, but now I'm being threatened again. Of course, divorce creates animosity between exes, but I never expected my child's father to sabotage his own child in order to make my life more difficult. The one who matters is our child, not me.

Using his lawyer as a proxy, Cliff continues to batter me in and out of court. I thought protecting my child would be seen as brave—instead, I'm painted as irrational, and it's soul-wrenching. We never have a trial. Like many custody cases in the 1990s, especially in Santa Clara County's overburdened family court system, everything plays out through a drawn-out series of hearings—confusing, procedural, and disorienting. There is no single moment to make my case, no structured presentation of evidence. Just a slow bleed of credibility, shaped by a mediator's biased reports and impressions. The only time I am allowed to speak directly about my experience is during mediation with the court-appointed evaluator. She takes an immediate dislike to me. Yes, I am emotional—often on the verge of tears—and she notes that as a deficit. Meanwhile, Cliff comes across as affable, easygoing, unflappable. That word, *affable*, follows him through every document, while mine is *emotional*—as if tears disqualify me from being rational. Since we never reach trial, accusations take the place of facts, and the courtroom becomes a stage for innuendo rather than truth. It seems clear that custody will be decided not by evidence but by the appearance of composure. And in that particular system, being overwhelmed is mistaken for being unstable. Trying to protect my child should have made me a hero. Instead, it made me a target. I am accused, second-guessed, disbelieved. The heartbreak isn't just personal—it's existential. If a mother can't protect her child without being punished for it, what kind of world are we living in?

There's a balloon effect happening. As I succeed professionally, I take two steps backward personally. In the height of the dot-com days, Wordcasters is one of very few companies that is both highly profitable and deeply innovative, but the court case has become a growing source of alarm. My steering committee colleague from Webgrrls, Jennifer Hughes, has stepped in part-time to handle PR and administration for Wordcasters, which is getting attention from the world's leading technology firms and top Silicon Valley venture capitalists. Jennifer helps me keep some of my balls in the air as I attempt to run a high-tech start-up while managing this family drama behind the scenes.

Creating an internet company from scratch, building the core product, maintaining servers and systems, setting up clients, and performing the work that my technology enables is only the beginning. Owning and running a business requires overseeing and motivating contractors, lawyers, accountants, and programmers; inventing the follow-on products to keep my lead in the marketplace; holding the interest of venture capitalists who have their eye on me; staying relevant in relationships with investors current and future; attending networking events, including Webgrrls; and, of course, making sure my car doesn't get towed from its metered spot below my window when I decide to drive into the city.

I know, deep down, that I'm at risk of being pulled away from either my child or my business. If I don't choose between them, I sense that the decision will be made for me, and I will probably not like the outcome. It doesn't help that I'm looking for another lawyer to replace the first one. Once I figured out I was losing, I wanted to switch gears to meet the challenges I could see I had been unprepared for. One of the problems I created for myself was that I didn't take any additional salary from the company coffers to cover my legal bills—my justification was that this was a personal problem, and my investors should not have to pay for it. But the deeper truth was that I was hiding the struggles in my private life from my colleagues. That meant I didn't have the funds to hire the kind of sharp legal strategist I now know I needed. It was a self-inflicted wound—and by the time I admitted I was in over my head, no lawyer wanted to take on a losing battle, mid-case.

One day, Jennifer arrives while I'm on the phone with Cliff's lawyer, who is skilled at verbal bullying. She tells me there's an OCR and I need to file a notice for FRC or there will be an ex parte—and she still has not received my IED and SAD, which she demanded in court last week.

"What are you saying? I don't know what any of these letters mean . . ." I protest.

Even my studies in legal terminology at court reporting school did not prepare me for this level of legalese. I know she's trying to confuse me and knock me off guard.

She cuts me off. "If you continue to waste my time, my client will have no choice but to apply for an MSC and seek compensation for all attorney's fees and expenses."

"But I . . ."

"I will have no more of your delays. We will see you in court!"

I'm learning that justice, at least in the systems I've encountered, is more performance than principle. I went in expecting fairness, expecting that if I just told the truth clearly enough, someone would act on it. But I'm finding that truth without power is often ignored—or worse, weaponized against you. What keeps me going is my own sense of what is right for Emma and my family. If I let go of that, I'll have nothing left.

The attorney hangs up on me. As I stare at the phone in my hand, wondering what just happened, Jen tries to get my attention.

Her face is flushed, and she's holding the other conference phone with her hand over the receiver. "You need to take this call! It's *The Financial Times!*" she mouths.

She puts the caller on hold and gives me a thirty-second PR briefing.

"Okay, this is *big!*" she says. "The reporter watched your livestream of the Voice on the Net conference in Boston, and they are doing a piece on you and Wordcasters. Give him some quotes they can use in the paper. Don't tell him too much, but keep him on the line and interested. This is going everywhere. It's international!"

I pick up my phone and press the blinking extension as Jennifer sits on her hands, smiling supportively. Over the next twenty minutes in conversation with the reporter, Tim Jackson, I feel remarkably calm and get out some sexy quotes without accidentally divulging any trade secrets to my competition.

When it's over and I set the receiver in its cradle, Jennifer jumps up from her chair and screams, *"THE FINANCIAL TIMES!"*

I've never heard of them, but for her, I pretend.

A week later, I walk down to the newsstand on the corner of Market and Front Street and buy a stack of the pink newspapers. I have three full columns, which I start reading right there on the sidewalk. I'm upset about the things they got wrong about my invention, worried about the things they got right in case they clue in my competitors, and highly pleased to see the words "brilliant," "impressive," and "proof of the technology." One of the world's most respected business papers, *The Financial Times* of London, has acknowledged that I am the only one doing livestreaming text broadcast on the internet. I'm thrilled by the optimistic last line: "Quinn is seeking venture capital this summer. She is likely to have many takers."

Hope came in strange forms back then—a working prototype, a client callback, a loan approved. I clung to each sign like a breadcrumb trail out of the woods. If I could make this one thing work, I thought, maybe everything else would click into place. That drive to create gave my days meaning and shape when nothing else made sense.

The city looks beautiful today. I float back to my office on air, smiling at everyone on the street, the pack of pink papers under my arm like a delicious secret. The phone starts ringing off the hook with inquiries. Jennifer was right. This is big.

CHAPTER EIGHTEEN

THE NIGHT WE FLIPPED THE SCRIPT

"For most of history, Anonymous was a woman."

Virginia Woolf

1998—SAN FRANCISCO, CALIFORNIA

On Wednesday, January 28, 1998, we pack over three hundred people into a ballroom built for two hundred at the generously donated Sir Francis Drake Hotel in San Francisco for the first-ever Top 25 Women on the Web awards. The ballroom is over capacity. With all the seats taken, women line the walls and sit shoulder-to-shoulder—on the carpet, pressed into corners, and perched on the edges of decorative planters. I cofounded the event through San Francisco Webgrrls because mainstream tech rankings—like *Time* magazine's "CyberÉlite" list—kept ignoring us. They honored Lara Croft, a fictional video game character, while the actual women building the web were nowhere to be found. As journalist Amy Moon pointed out in her *SFGate* coverage after the event, "What does it say about the importance of women in the industry when, on the opening page of the article, the three "cyberélite" pictured are Larry Ellison, Paul Allen, and Laura Croft?" When the industry's idea of a "top woman in tech" leads with two billionaire men and a cartoon with large breasts wearing nothing but a utility belt with pistol holsters and a leotard, something is seriously broken. So we made our own list.

It wasn't about celebrity status or IPOs. It was about who was actually doing the work. We put out a call for nominations through the international Webgrrls network and received over one hundred submissions—peer-nominated women making real impacts in engineering, publishing, accessibility, community building, design, and start-up leadership. Our judging panel—Valerie Hoecke, Linda Robertson, and Megan Taylor—narrowed it to twenty-five based on vision, substance, and influence.

I was still scrambling until the doors opened—laying out fruit, cheese, juicy meatballs, and focaccia bread so the women coming straight from work had something real to eat. It was what I did every month as founding steering crew member of Webgrrls SF. Elaine Sosa, our chapter leader, secured the wine donation from Heublein and kicks off the night with grit and charm: "We even let the guys come. Heck, some of them even know something. We'll drink a beer with them, pick their brains."

The vibe is electric. This isn't a corporate-sponsored industry mixer—it's a reset. Eleven of the honorees in the room are each presented with a yellow rose and a rolled decree of Webgrrls excellence before they take a seat at the long table on stage. When they speak, they speak for all to hear.

Laura Lemay tells the story of seeing a random USENET post asking for writers to cover HTML. She applied, learned HTML over a weekend, faked having an outline, and got offered a book deal—on the same day she was offered a job as employee number twenty-five at Netscape. She took the book deal. "Say what you're good at," she tells us. "Say it clearly. Say it with confidence."

Brenda Laurel, VP of design at Purple Moon, cuts through the buzzwords: "It's almost never about technology. Reach through it to what it is about: people, about culture, about relationships." She adds, "Keep in mind who you're working for, who you want to move and change. And if you manage others, ask them what they think."

Sarah Allen, an engineer at Macromedia, makes the mission tactical: "Whenever there's a job opening, interview as many women as you can. In engineering, you really have to go find women to interview." She reminds us to look for mentors—not just those who are successful but those who are successful in ways that resonate with us.

Mary Furlong, CEO of Third Age Media and founder of SeniorNet, adds even more to the evening's theme: "It's not about the technology. It's about the people. Go to the centers of influence. Learn the vocabulary. Then teach them your words—trust, caring, reciprocity."

Janelle Brown, cofounder of *Maxi Magazine*, doesn't mince words: "I'm here because I'm loud and outspoken. One thing I've noticed is that women aren't very good at self-promotion. I get tons of press releases from fifteen-year-old boys in Kansas about their dorky websites. I never get any from girls."

Robin Wolaner, EVP at CNET, speaks directly to our value: "You gotta be pushy to be paid fairly. Look to where the growth is. Growth often ignores minority status." Then she says what many of us are thinking: "As women advance, we do right by other women—because we don't know how not to do that."

Cynthia Waddell, creator of the web page disability access design standard, couldn't attend—so her husband Tom accepts on her behalf. She's too sick to sit up, but her influence is in the room. Elaine nods toward him and says, "Gotta support your grrl." Someone in the crowd shouts, "Executive husband," and the whole room cheers.

The press is everywhere—*Wired, SFGate, Developer.com, E-Business*. Sheri Kramer from Macromedia says she's here to see who the good speakers are because her conference speaker lists are still dominated by men. Elizabeth D'Errico of Headland Digital is inspired to see the impact women are having in the tech industry, pointing to the power imbalance still in place in upper management.

Not everyone agrees. Kim Locklin, CEO of Foxmail, tells reporters she feels like an equal and that jobs are filled based on skill. I've heard that before. But I also see who gets profiled, who gets funded, and who gets left out of the narrative. It's not just about skills. It's about visibility. That's what we're correcting.

Even some of the men are caught off guard. PF.Magic's Adam Frank and Richard Lachman, video game developers of virtual "Petz," showed up expecting something more generic and less female-oriented. But they stay, seeing the potential in the developments of the panelists.

Craig Newmark, Craigslist founder and frequent supporter of Webgrrls events, greets me with an air kiss.

A woman I've never met but whom I recognize as a force of nature, Tiffany Shlain, is here too. She pitches me about launching her own new concept of tech awards—The Webbies. Something is shifting. It seems like every rising star and established name has found their way here.

By the end of the night, the room is loud with conversation, clinking glasses, and the kind of laughter that signals relief, not just joy. We're not competing for one seat anymore. We're building a bigger table.

This night is a blueprint of what the world will look like when run by an equal share of women and men in leadership. It becomes an annual tradition—but tonight, it's a marker. We're showing the next wave of what's possible. As I was quoted in the press at the time, "It's important to women in our industry that the visionaries, leaders, and powerhouses— our role models—are thrust into view on a regular basis."

That night, we put them center stage.

All the while, I'm quietly fighting exhaustion. I've been sick all week, nauseated and shaky. Any other event, I'd have stayed home. I suspect I might be pregnant again—on top of the custody battle—but I push through anyway. Because that's what we do. We show up. We lead. We build what hasn't existed yet—and we make it impossible to ignore.

All quotes used in this chapter from:

Amy Moon, "Webgrrls: Top 25 women on the web," *SFGate,* January 30, 1998, https://www.sfgate.com/news/article/Webgrrls-Top-25-women-on-the-Web-3014827.php.

CHAPTER NINETEEN

PAPER TIGERS, APEX PREDATORS

"One by one the stars blink out, but I am learning the dark."

Ada Limón

1998—SAN JOSE, CALIFORNIA

Back at home, the widening imbalance between the professional success and the diminishing personal peace I've been experiencing for months now is heading for its breaking point. Before the custody battle gets completely out of control, I ask my mother and Sal to connect me with an attorney in San Jose. My hope is that Sal can bring Link back in, or another attorney with powerful chops. I realize how wrong I was to have chosen to go easy at the beginning of this court battle. I decided to take the money I needed to boost my salary, so I don't need funds—I need connections.

Sal is a high-ranking member of the San Jose Rotary Club, and, due to his position as an important building tycoon and now as an angel investor in Silicon Valley, he has a team of lawyers at his disposal at all times. He calls them fixers, and they help smooth out legal issues for him. When people close to him got into trouble, Sal leaned on his fixers to take care of things for him. Because of them, no one faced any real consequences for their actions.

I'm afraid to ask my parents for help, but I figure they'll be willing to do something for Emma, their first grandchild, whom they love so much. I call my mother, and she seems sympathetic.

"But I'll have to ask Sal," she says.

This is a well-worn phrase.

After a while, she calls me back.

"Sal wants you to meet him at his office in San Jose next week. You can present your case to him then."

I worry the entire week, but I hold on to a little bit of hope. On the somber drive down to San Jose, I decide to do whatever Sal tells me to do. I'm sure he'll help. He knows I'm a loving, caring, capable mother, and he doesn't want any harm to come to Emma.

I've forgotten how much Sal enjoys seeing me on my proverbial knees, groveling for his mercy. He sits behind his desk, perfectly sanctimonious, while I plead my case. On the credenza behind him, I notice, are framed photos of my mother and their two sons—no trace of me.

When I finish my speech, Sal begins his lecture. I am stupid, he reminds me. I got myself into this mess by choosing to leave my marriage. I've brought all of this on myself.

It's everything I've heard from him my whole life.

"You got yourself into this mess with your delusional optimism. You think too highly of yourself, like you always have. I've invested in a lot of start-ups, and most of them are a bust. I'm not about to support your stupid dreams that are alienating you from your own daughter. If you stayed home like a real woman, none of this would be happening to her."

I have an ace in my pocket, though.

During the last court hearing, Cliff's lawyer introduced a stronger angle to their narrative suggesting I am acting out due to repressed memory to the detriment of Emma. To prove that I'm an unfit mother guilty of "parental alienation," she argued that I was sexually abused by Sal as a child. That experience, she claimed, is why I now hate men and why I've turned Emma against her father.

"Susan may have repressed the memory of her own sexual abuse," Cliff's lawyer pronounced. "But that latent understanding led her to instruct Emma to claim that Cliff had sexually abused her."

In the mediation room, I sat flabbergasted. But how could I defend myself against a patent falsehood? I was abused but not sexually, and I do not hate men. They were creating a set of false facts—and claiming I didn't even remember them myself—to punish me for investigating Emma's unsettling statements.

In his office, I explain to Sal what happened with the case. Now he has a personal reason to get involved.

"Before I decide to help you, I'll have to talk to your mother," he finally says. "I'll let you know."

Several days later, I still haven't received a response. Too anxious and desperate to wait any longer, I call their house.

Sal answers the phone, but he makes a quick excuse.

"I've given your mother final say this time," he says. "If she wants to help you, I'll go along with it."

Then he hands her the phone. My hopes rise. Surely my mother will choose to help us!

When she takes the phone, my mother sounds exhausted, distant.

"We had our lawyer threaten Cliff and his attorney with a counter-suit if they do not immediately retract their slanderous accusations against Sal," she explains. "Cliff's lawyer agreed to drop it."

I wait breathlessly on the other end of the line.

"Beyond that, you're on your own," she says. She wishes me luck and tells me she hopes I won't lose Emma, then hangs up the phone.

I am dumbfounded and devastated, too full of sorrow to even be angry. My parents saved themselves but left us stranded. It would have been so easy for their lawyer to throw some muscle in our favor. It would have been so easy for them to offer us a lifeline. Throughout the custody battle, I've been hoping, stubbornly, that they would see me flailing and reach out. Maybe it's the child in me, just wishing her parents would see her and love her the way they're supposed to. But they keep proving me wrong. I can think of a thousand ways they could've shown up for me,

and each time, they didn't. If not for me, why not for their grandchil-
dren? I wanted my kids to have a relationship with their grandparents
like other kids did. That grief, now extended to grief I feel on behalf of
my children, is more devastating than when it was just me they hurt.

"You always want more," my mother told me once. "Nothing is ever
good enough for you, Susan."

Now, I'm nearing my breaking point. My success seems to have made
me a target. Sure, I wasn't bashful about sharing my wins with my family, expecting their acceptance and admiration. In hindsight, I should have
hidden my meteoric rise or downplayed it instead of rushing to brag about
myself. Maybe if I had been more humble, Cliff wouldn't have focused on
me so much. Maybe flexing my muscle without a care for who saw me
winning—my blatant taunting, unabashedly sharing my own perceived excellence—made me, and therefore my children, a target. I'm still looking
over my shoulder. I don't know how to inhabit success. I only know how
to chase it.

Somehow, over the next month, I hire a new attorney without my parents' help, but only on a consulting basis. He's unenthusiastic about coming
into the case midway, but at least he knows the meaning of the acronyms
and initialisms I've been struggling with. I don't get calls from Cliff's attorney anymore. With my new lawyer's help, we cobble together the bones of
an agreement. It's not perfect, but it's something I can live with. I will retain
physical custody. Another hearing date is set, then arrives.

In the courthouse bathroom, I splash my face with water. In the mirror, I find exhaustion so obviously casting shadows across my face. While
all this has been going on, I confirmed that I am indeed pregnant again,
and the morning sickness that lasts all day has begun, just as it did when I
was carrying Emma and Olivia. The custody case, my business, protecting
Emma, and now the pregnancy—everything is happening at once. I wish
I could just press pause. If I could shrink myself enough to slip down the
drain and be anywhere but here, I might. But I must cross this last hur-
dle for Emma. We're so close to the finish line. When this is through, life
can get back to normal, at least somewhat. I don't know how to live in this
conflicting reality. I'm fighting just to keep my daughter under my roof.

It's like I'm being punished for trying to keep her safe. What can I do now to protect her when she's not with me?

My new attorney has a scheduling conflict and can't attend today's hearing. I'm not too worried about that since all we are doing is filing the order. He tells me if anything goes wrong, I should ask for a continuance, and the judge will approve it.

"I understand both parties have reached an agreement," the judge booms.

I try not to look at Cliff on the other side of the courtroom.

"I'm afraid we have not, Your Honor," his lawyer answers.

What?

No, no, no, no, I think. This can't be happening. That no good, dirty, lying lawyer has gone back on our agreement. She starts spewing words I can't understand. The room is collapsing in on me. I can hardly catch my breath.

"At the advice of my attorney, I'd like to ask for a continuance," I say into the microphone.

Cliff's lawyer objects, accusing me of manufacturing a delay and wasting the court's time.

The judge is visibly annoyed.

Cliff's lawyer says, "The petitioner is ready to go to trial now."

"Then I suggest you all go out into the hallway and work something out," he says. "When you come back in, I expect you to have a signed agreement."

My feet are concrete blocks beneath my knees, but somehow, they carry me to the hallway. It's a bustle of people moving in every direction, their shoes clicking, stomping, squeaking against the slick stone floors to create an uncomfortable symphony. Every sound is magnified and reverberates inside me, knocking around against my teeth. I'm so tired.

Cliff's lawyer approaches me in the hallway, and there is a new set of papers in my hands now, tiny, typed legalese text blackening the pages stuffed inside a manila folder. I blink at the sentences. Around me, the world spins and spins, gaining velocity with every rotation.

There isn't time to make sense of anything. One sentence becomes another on the page. I'm so tired of being beaten up.

My only relief is Nancy, Jake's lawyer sister, who has volunteered to accompany me today since my lawyer could not. Surely she can make sense of this jumbled jargon. I don't read it. I can't.

"Susan," she says in the hallway, "I think you should sign."

Emma is at home with a neighbor, waiting for me. It's a crisp spring morning. Olivia is waiting too. The judge is waiting.

Nancy hands me a pen. It inks my signature so smoothly onto the page, unaware of its crime.

Cliff and I will have joint physical and legal custody of Emma. His time with her will dramatically increase to include almost all her time outside of school days. I am responsible for delivering her to him every Wednesday and every other weekend, a two-hour drive each way. She will spend Christmas, Thanksgiving, and Father's Day with him, plus both of their birthdays and two months in the summer. Within seventy-two hours, I must surrender Emma for a two-week visitation with Cliff. Then, after everything, I will reimburse Cliff for all the attorney fees he has accrued since this custody battle began two years ago. The betrayal doesn't just wound me; it reorders my faith in systems and in people.

I am too overwhelmed by the legal jargon to grasp, at first, just how punishing this agreement really is. How am I supposed to run a company when I'm constantly ferrying my daughter back and forth? How am I supposed to be an effective parent when I can't even take my daughter out of the county without her father's express permission? It's not coparenting— it's surveillance. Cliff receives nearly all non-school time. I become his courier, required to deliver Emma midweek and on alternating weekends. He controls holidays, birthdays, even geographic movement. And if I fail to meet any condition—even accidentally—I forfeit all rights. Outside the courthouse, I spy Cliff and his lawyer kissing and celebrating. I still can't process the breadth of what they've won on this day when I showed up weak, exhausted, and ill-prepared.

The drive back to Mill Valley is long and quiet. Jake is driving, and Nancy is in the car with us—but I'm somewhere else.

At home, finally, I walk, distracted, past the children. I paw my way to the bedroom, loosen the constricted muscles of my trachea, and unleash a carnal scream. I don't have awareness of or care who hears it. The neighbors. My family. Wails spill from my lips one after another, the sound of a lion dying in the jungle. I don't know where these sounds are coming from. I understand, yes, they are coming out of my body, but they feel like they were born from the molten center of the Earth and traveled up through me and out into the sky, over the oceans of the world, past the moon and beyond the stars into the darkness of deep space.

With each wail, I release all the pain and sadness I've been carrying, the betrayal and the disappointment and the wreckage of our lives. I fill the room with my shame, too immense to stay inside my one human body. Shame at my own stupidity, my own weakness. Shame at signing away so much of my daughter. At not being strong enough in that crucial moment to go on fighting for her. At thinking I could have my companies and my family and my freedom from Cliff and his control. Shame at believing in myself and for letting my child suffer so dearly. I thought I was shielding her, trying to take all the hits myself. I didn't see that she was standing right next to me, getting hit too.

A child should never have to navigate that kind of anguish alone. When adults prioritize control over care and denial over truth, they don't just fail a child—they force her to carry their failures as her own.

Eventually the wailing stops. I am emptied of the sound. Through the window, I can see the sky turning pink as the sun slides toward the edge of the sea, where it sets out of view. I have survived this day.

My mouth feels dry, but I can't think about that now. I am hurtling toward a decision, perhaps the most important decision of my life.

I have to shut down my company. I need to save Emma. I need to find a way out somehow. We have to run. We have to escape this oppressive man. Now that the custody case is over, I can address the real problem. The one Cliff has been trying to obfuscate with false allegations and legal maneuvers, all constructed to conceal his own crimes. We will run, and Emma will be safe.

I close my eyes and remember the panic and the excitement I felt walking through the doors of the Microsoft headquarters. The truth is that with my work, I'm on the precipice of something incredible. I've forged so many inroads and overcome so many obstacles, all to pioneer technology that changes the way information is accessed and understood around the world. My brain is filled with ideas, new avenues to take, new ways to innovate, expand, explore. The potential is infinite.

But, in this moment, none of that matters.

Not Microsoft. Not Mark Cuban. Not one scrap of hardware or software. Not one measly dollar in the bank. If I shut it down now, there may be a path forward. For my daughter. For my family. For me, as a mother, as the person I want to be most of all.

As I wailed, I grieved. Not just the loss of my business, but the loss of my belief. I had poured so much into making this work. I had survived things that should have broken me. And still, it wasn't enough. If even this isn't sustainable, what is? I don't have an answer. But I know I have to keep showing up for my kids. That becomes my new baseline.

1983—SAN JOSE, CALIFORNIA

Six months after my first runaway attempt, I try again.

The usual pattern repeats. I run, am chased, make it to the door. The door locks behind me. But this time, I have a Velcro wallet hidden in the side yard by the trash cans. It has about ten dollars in it, enough to get me somewhere. I sneak around the side of the house, quiet enough not to be heard, collect the stashed wallet, and get myself to the bus stop. From there, I take the city bus to my friend Beth Hanson's house near the Gunderson High School in Almaden Valley.

At Beth's, I tell them everything, and her family gets down to business. They are not taking any crap. They are not intimidated! Beth's dad gets involved, and conversations happen over the phone that I am not a part of.

The adults listen to me when I say, *I can't go back. I'm never going back. Not this time.* Beth's father looks at me soulfully without saying a word. He sees the truth in my full, round eyes.

"Okay," he says, almost to himself.

The Hanson parents send Beth and me up to her room to hang out and relax as they set plans in motion. It must be quite a feat they're performing downstairs.

The Hanson house is a fun, rock 'n' roll kind of house. Beth and I listen to Boston and Bob Seger on the stereo in her older brother's room. We play "Hitch a Ride" again and again, real, real loud. The lyrics and the sound of the sonorous electric guitars and synthesizers comfort me and give me hope.

I stay with Beth and her family for the weekend. On Sunday afternoon, a phone call comes for me.

It's my grandfather, Baba. He doesn't comfort me or engage in small talk. Instead, he needs to evaluate me. Am I a troubled teen making up stories? Am I a pawn for my parents to find my real father so they can sue for child support or send him to jail?

The day my father gave me the gold watch was the last time I saw my dad.

Six years later, there is still no sign of them. Until now.

"We're not entirely sure yet," Beth's parents told me, "but we may have found him."

"I heard you're having trouble with your parents," Baba says, his voice hauntingly familiar through the telephone. I can almost smell the once familiar scent of his pipe tobacco. "Is your stepdad hurting you?"

"Yes," I answer. "That's true."

"How would you feel if you knew your father lived in another country? Somewhere not as nice as here. No luxuries or anything like that. Honduras. Would you want to live in that kind of place?"

"Yes," I answer without any hesitation.

I don't know where Honduras is, but it sounds a little dangerous the way he is describing it. It doesn't matter, though. Any place is better than here.

It's clear from Baba's vague questions that he's trying to protect my father. *Whose job is it to protect me?* I wonder.

I have a sudden wave of feelings—fear, sadness, a little hope, and . . . possibility? Some truth coming to light? I feel like I am being given the chance to go through the hidden door or down the path I've been dreaming of for so long that it's more like a myth than reality. I might actually see my father again. Could this be real? I start to piece together the truth of it. I'm steps away from my father, my real father. The reality of it is so unbelievable that I'm actually afraid. I'm afraid to leave the comfortable mystery of him.

With the help of these caring adults, I've opened the door to a new life, just like that. I feel an overwhelming desire to draw back. My brain is screaming, *Warning! This is too good to be true!*

The next call comes late Sunday night.

"Hello? Susie?" the voice crackles through the line. "This is your dad."

His calm voice eases through the static. He sounds laid back, warm, welcoming, not at all suspicious of me like my grandfather was. His voice feels the way his arms would, wrapping around me and lifting me up.

"Daddy?"

My mind bubble wraps around this. Is this my Daddy? This isn't a dream? Yes, it is actually happening. I burst into tears.

"I . . . can't . . . talk . . ." I choke on my sobs, unable to see through my tears.

I am in the Hansons' kitchen on their yellow phone with the long curly cord. Beth's parents are somewhere nearby, giving me privacy but also there in case I need them. I'm quite sure Beth's tough, bearded, Bob Seger–loving dad is crying too.

"Oh, Susie, it's okay. Let me do the talking," my dad says, calmly, as if this is the most natural conversation in the world.

He tells me he heard about my situation and that he has already spoken to my mother.

"Do you want to come live with me?" he asks.

I choke out the one necessary word: "*Yes!*"

"Okay, honey, don't cry. Your grandfather will make the arrangements, and he will bring you here to me. It's not fancy. We live on an island. We don't have much . . ."

"It's okay," I say. I swallow my tears and really try to listen and absorb all the information in case it does turn out to be a dream.

"Okay, honey, I'll see you soon."

Two whirlwind weeks later, I am on a flight with Baba. I've never been on an airplane before. We arrive in Honolulu—not Honduras. It turns out that was just a trick to test my resolve. When we land and the plane door opens, the air pours in hot and wet. I've never been anywhere like this. I breathe and blink up at the hazy night sky.

The Esaus, a Samoan family and friends of my father's, are waiting outside. I collect my suitcase, and we pile into their large station wagon. Since he's only a chaperone, my grandfather gives me a perfunctory good-bye and leaves me to my fate. He has an early flight back to California in the morning.

The Esaus drive me along the beach of Waikiki. I'm exhausted, but I rouse myself enough to look at the ocean on a Hawaiian beach for the very first time. I can't believe it. *This is my new life?*

That first night in Honolulu, I'm paralyzed by the shock of so many firsts. The Esau family lives in a skyscraper, and I've never set foot in such a tall building before. Strangely, they leave all the doors open at night. In the front room overlooking downtown Honolulu, I can see other tenants passing by the open front door from time to time. Though January, it's very hot. I've never really experienced humidity before, and my skin feels wet even though it isn't. My hair somehow hangs longer than it did at home, and breathing feels thicker.

The plan is to stay in Honolulu tonight, then I'll fly alone to the Big Island in the morning. It's weird spending this time with my father's friends. I don't know my father, so naturally, I don't know these people either. I try to imagine how they know each other. Bernie Esau is the mother of five kids, all teenagers and mostly boys, except for her daughter, Maliki Esau, who is a couple of years younger than I am. Bernie feeds me spaghetti from a big pot on the stove. As I eat, the family eyes me curiously.

With a forkful of pasta in my mouth, I suddenly freeze. I chew and try to swallow, but the masticated noodles in my mouth start to taste like burnt human flesh or rancid meat, undigestible. I panic, not wanting to insult my hosts, but my body is desperate to expel the food. *This isn't normal*, I think. My senses are clearly out of order. I run into the bathroom, my mouth still full, and spit its contents into the toilet before becoming sick. I am sweating profusely as fear and shame overcome me. Why can't I control my behavior? How can I be so rude to these lovely people who are helping me?

The family is speaking English, but I cannot understand them. I realize later that it's pidgin English they're speaking, a dialect of Hawaii that I will come to understand easily before the year is out. Perhaps my age saves me from total humiliation. I am sixteen years old but have the delayed maturity of a much younger kid. They must feel sorry for me and understand that my rudeness isn't borne from disdain. I'm just a terrible guest. At the family's insistence, I spend the night in Maliki's bed while Maliki sleeps on the floor next to me.

In her room, Maliki barely speaks to me, but her curious eyes follow my every move. She expresses excitement when I pull out my Sony Walkman, so I pass it to her. She turns it over in her hands a few times, not knowing what to do with it. I show her how to put the headphones over her ears and hit the play button. Her eyes leap up to mine, and her smile explodes as she hears Duran Duran's "Union of the Snake" through headphones for the first time. I think back on my first Walkman experience a year or so ago, that feeling of possibility, magic, autonomy. Control of my own music, a private concert in a way.

I let Maliki wear the Walkman as we both fall asleep. Of course, I want it too—to escape into the familiar blackness of the music in my ears—but I don't have the heart to take it from her. Her beautiful face, full of curiosity about me and my exotic clothing and shoes, is too sweet to disturb. Her smile tells me I'm among friends—even if they have different foods, different smells, and a different language.

In the morning, I am taken back to the airport for the short flight to the Big Island, where my father and my brother Bryce live. Nervous, excited, I look out the window, almost dying of anticipation. I'm an hour away from seeing my dad! My expectations are so high—so, so, so high. This is a dream come true, and somehow, it's really happening. I can see out the window that I am in Hawaii in the daylight. This is real. I'm here.

The first time I see my dad, he is standing at the bottom of an escalator in the open airport lobby with my little brother Bryce, who is now seven. They both look up at me as I descend. My dad's smile is beaming. He is holding a bunch of flowers that I'll learn is called a lei, and as I walk off the escalator, he drapes it around my neck. The blossoms are beautiful, fresh, fragrant. Then, he leads me to the car. I am at a total loss for words.

We arrive at a beautiful house on Waianuenue Avenue. It's surrounded by an incredible garden and a large wraparound porch. Birds and flowers are everywhere. On the porch is a tray of mangos, pineapple, and sliced ham. I can't believe this isn't a dream. The colors that envelop me are surreal.

I am, quite simply, in shock. Food still tastes like poison. My father hands me a piece of guava, but the fruit burns my tongue. I can't manage to swallow it and instead spit it out into a napkin. My most basic senses are all messed up. The shock lasts two weeks before I finally start to emerge from it.

I enroll in the eleventh grade down the street at Hilo High School, and I take the yellow school bus in the mornings with my brother. The bus stops right in front of our house, then picks up all the other kids on its route, zigzagging around Pe'epe'e Falls.

My very first day, a girl catches a glimpse of me, the bewildered new girl, as I climb onto the bus and find a seat. Within seconds, she plops down right next to me, her bookbag bouncing as she sits. She seems to deliver an entire speech before she even sets her bag down on the floor.

"I'm Nani Ball. Pleased to make your acquaintance," she says, thrusting out her hand to shake mine.

I shake it.

"What are you, a sophomore? Where are you from? Wait, let me guess. Cali-*fornia?*" She says the name of the nearest state as though it's some exotic location, making fun of it but in an honest, good-natured way. "What's your name?"

"Susan," I say, wide eyed. It's my only contribution to the conversation.

For the rest of the day, I watch and listen, following on autopilot. Slowly, the fog I've been living in starts to lift. No one here knows that I've just been reunited with my father and brother, who went missing six years ago. No one knows that I'm escaping a dangerous home life back in California.

"I'll take you to the corner where we sit, and then you'll need to get your class schedule. I'll show you where to do that. Don't worry, you can leave all your stuff in our corner," Nani prattles on.

I'm so grateful for her. Having a friend can do wonders for a teenage girl. Soon, I begin to fit in. I become a good student, easily earning an "A" in every class. Maybe that's because I'm finally safe. I'm allowed to be a child. I'm allowed to have a life—for the first time in so many years. I look at the world with my own eyes. It is miraculous.

Miraculous. A miracle.

CHAPTER TWENTY

SPELLBOUND

*"No one has ever written, painted, sculpted, modeled, built,
or invented except literally to get out of hell."*

Antonin Artaud

1998—MARIN HEADLANDS, CALIFORNIA

"**D**oes everyone know 'We All Come from the Goddess'?" Zsuzsanna Budapest shouts over the sound of the rolling waves, her Hungarian accent slipping out.

It's late July, and I'm eight months pregnant with my third child. Our closest friends and family stand in groups on the otherwise empty Muir Beach under the shadow of the majestic Mount Tamalpais. They're all here for our wedding, officiated by the famous witch Z. Budapest. Jake's buddies are costumed in garish Hawaiian shirts and oversized sombreros, and my friends look lovely in sundresses. Everyone is exchanging funny looks. Is this actually happening?

Our last-minute beach wedding is a mix of fantasy and disbelief. What's clear, however, is that none of us knows "We All Come from the Goddess."

"Is it a song?" someone shouts.

Zsuzsanna is momentarily frustrated. I've been following her teachings in my search for meaning and power during the tumult of recent events, and I have read a few of her books, including *The Goddess in the Office*. When I found out she lives just across the Bay in Berkeley, I somehow recruited her to officiate. Might as well include something authentic in this fake wedding I've cooked up as part of our escape plan.

"Okay, everyone!" she shouts. "What song do you all know?"

The crowd mumbles, unsure. The guests seem to wonder who is in charge here and what is going on. Jake and I are at the head of the group, but we're letting Zsuzsanna run the show. Except for her, it's everyone else's first witch wedding at the beach, after all.

Finally, one of Jake's friends volunteers "Yellow Submarine," and the crowd seems to agree.

Zsuzsanna shakes her head. "Okay," she says. "We sing this song and follow the one with the broom."

My bridesmaids, Tia, Marla, and Ramona, have decorated the witch's broom with flowers. Marla holds it high over her head and walks toward the water. The rest of us follow, singing.

In short order, it becomes clear that only Randy in his oversized Mexican sombrero actually knows all the lyrics. The rest of us only know its chorus. Regardless, Randy belts out the words. A few guys in their Hawaiian shirts provide bugle noises, their lips slicked by the rum punch they brought in an ice cooler.

It isn't an official, recorded wedding, but with this, I've found a loophole in case law that will allow me to "follow my husband" out of the Bay Area for a job. It's our best chance to escape Cliff by creating some essential distance between us. This was all I could do within the limits set by the courts.

Once I dissolved my business, Cliff seemed satisfied that I had fallen so far and that his punishment had worked. After fighting so hard for parenting time, he even began to give up his additional visitations. He didn't want Emma; he wanted to hurt me—and presumably save himself from any investigations that might arise should Emma continue saying uncomfortable things about him.

Under the agreement, I was required by law to reside in the greater Bay Area, but we fell in love with some land by the beach in Humboldt County. Land in the country! With chickens and the ocean and farming and cheap rent! It'd be the perfect place for us and the kids—exactly what I started my business to earn for us. But it was beyond the border of acceptability according to our new custody agreement.

Luckily, I found some case law allowing a parent to legally break their arrangement if their spouse found work in another area. I decided to take a risk on that basis. Jake easily found a job at the Eel River Sawmill in Fortuna, but he wasn't technically my spouse. As important as it was to ensure the safety of my children and the sanity of our family, I really did not want to get married again, especially if only to satisfy an oppressive custody ruling.

Thank heavens for California. With more research, I learned that courts in the state recognize a marriage as long as some sort of ceremony is performed in the presence of twenty or more family members and friends. In no time, Jake and I are on the beach with the people we love most and a famous feminist author and eccentric Wiccan who joins us in unofficial matrimony.

And the risk pays off! Jake lands the job at the Eel River Sawmill in Humboldt County, the judge never asks for a marriage certificate, and we move to Ferndale three days after the ceremony.

1998—MILL VALLEY, CALIFORNIA

A week before that beach wedding, I had a Transient Ischemic Attack, also known as a TIA or a "mini-stroke," while seven and a half months pregnant. Our neighbor and dear friend Tia came to pick up the kids that Saturday for a trip to the zoo, and the bright shine from her windshield caught my eye as she pulled into our driveway. Then one eye went dark. I saw a flash of light, stars, then blackness.

This is temporary, I thought. Like when you lose your vision for a split second after a camera's flash. But as Jake and I headed off on a trip to Costco for bags, boxes, and cleaning supplies to prepare for our move, I realized that my vision in that eye had not returned. We zoomed along Highway 101, and as we neared the store, I noticed too that my right hand and arm were going numb. I held up my arm in front of me and marveled at the fact that I could not feel it. Instead, I had a light and airy, detached feeling. I was not really concerned yet—merely interested in the strange phenomenon.

At Costco, I wandered away from Jake. We had a lot to buy in preparation for our upcoming move, so, to save time, I beelined to the paper towels while Jake maneuvered the large flatbed cart up another aisle. I picked out the towels I wanted, then headed back to meet Jake, but as I did, something strange happened. I looked around at the other shoppers, and it hit me.

I didn't know what Jake looked like.

I knew this didn't make sense. How could I not know what he looked like? I *must* have known what he looked like. But I didn't. I blinked wildly at the faces passing by me. Who was I even looking for?

Oh, I know, I thought, snapping out of the fear that was quickly creeping in. *He's pushing an orange flatbed cart*. Most shoppers had regular shopping carts, so I figured the orange flatbed would clue me in to Jake. I may not have known how to look for him, but once I saw him, I was sure to recognize him. Right?

Minutes passed. I stood in the center aisle, very pregnant, stupidly clinging to a pack of eighteen paper towel rolls. Eventually, I decided to abandon them. I started to feel very weak, and a bad taste, like tar fumes, filled my mouth. I was too self-conscious to ask for help even while thinking, *I need help! Why do I find it impossible to ask for help?* I tried to act as inconspicuous as I could until I could find somewhere to sit down.

Jake will find me, I thought. *I'll sit on a bench near the checkout lanes, and he'll find me.*

I worried he wouldn't see me there, but I had no choice but to sit. I was declining fast. As I stumbled toward the benches, I felt a wave of guilt at leaving the paper towels in the middle of the aisle.

After a while, Jake did find me. He had already paid for all the cleaning supplies we were here to buy, and he pushed the cart up to me.

"I need food," I gasped when I realized it was him. "I really need to eat. I'm feeling weak."

He bought me a hot dog, which I plunged into my eager mouth. As soon as it hit my tongue, it turned into rotten flesh. I spit it out.

"It's inedible," I told Jake. "I need to get food fast."

I was pregnant, so he probably assumed this was a blood sugar thing. Nourishment would correct the problem. He drove into a fast-food line and ordered something for me.

"I still can't see, and my right side is almost completely numb now," I said as we waited in the queue.

The cars in front of us inched forward. Suddenly, realization swept over me, and I was desperate to get the words out.

"I need to go to the hospital. Now!"

With that, Jake pulled out of the line and rushed me to the hospital. Later, he told me I spoke gibberish to him all the way there, like I was speaking in tongues.

We arrived at the emergency room at Kaiser Marin in San Rafael, and I was admitted immediately. The right side of my face drooped, as did my tongue, so I was rushed into testing. A CT scan and spinal tap later, we waited for test results. Jake called Tia and asked her to hang onto the girls for a while longer.

In my ER bed, tubes running every which way in and out of my body, I slowly regained my awareness. Once I was a few hours out of the fog, I was released—with an appointment for an MRI next week.

If the midwives find out about this, they won't let me have a home birth, I worried. What I absolutely did not want was another hospital birth.

Tia agreed to keep the girls overnight, but I wanted them with us. Jake drove me home.

"How are you feeling?" he asked, helping me into bed. "Do you think you're well enough for me to go into the city as planned?"

I put some pasta on the stove and pull out an assortment of cheeses from the fridge. It was a mac and cheese kind of night.

He had a bachelor party scheduled with all his buddies. *Oh, right,* I thought, *we're getting married.* It wouldn't be a legal marriage, but he was still keen on a night out with his longtime friends.

"Go," I told him.

But deep down, I was *not* happy about it. I was carrying our second child—and I'd just had a *stroke.* Shouldn't he have wanted to stay with me? Sure, I'd probably be fine without him, but I wanted him to pamper me. I wanted him to love me.

I spent a quiet night alone with the girls. Emma made me toast. She buttered it like a child would—more butter than bread. I ate every bite. She whispered to me, "Don't worry—I'm little, but I can be in charge sometimes, Mommy."

The girls didn't know anything about the hospital or the stroke, but they made a point to sit near me that night, cuddle a little closer, and that was enough.

In my solitude after putting the girls to bed that night, I reflected on the many times I've been abandoned, abused, or neglected. These normalized behaviors in my childhood helped me expect the same as an adult. *At what point,* I wondered, *do I stop blaming others for the way they treat me and start inspecting myself?*

Should I have asked Jake to stay with me? Why did I feel I *had* to let him go to the bachelor party? Did I rely on others to show me how I should be treated? How did I teach people how to treat me when I didn't know how to be loved?

I needed to understand that before I could find what I was looking for outside myself. For so long, I'd been focused on surviving—escaping danger, supporting myself and my children, building my company, battling in court. Maybe it was finally time to focus on healing.

CHAPTER TWENTY-ONE

A FINAL GAMBIT

"She was fire and ice, but most of all, she was tired."

Nikita Gill

1998—FERNDALE, CALIFORNIA

Once again, life throws me a curveball.

"Mama, you need to push!" the midwife screams.

At 8:30 that morning, before any signs of labor pains, Emma boarded the school bus in front of our ranch house on Port Kenyon Road in Ferndale. We'd be welcoming her brother to the family less than two hours later.

Not long into the birth, it's clear something is wrong. The midwife wears a headphone piece attached to the heart monitor. She takes it out of her ear and hands it to me to pop into mine.

Bubum, bubum, bubum.

"Mama, do you hear that?"

The beat is slow, steady.

"That's not your heartbeat," the midwife says solemnly. "That's the baby's."

Her words crystallize in my mind. I knew from all the maternity check-ups I'd had over three pregnancies that a baby's heart rate is fast, like a bird. My baby is failing.

Babum, babum. Too slow.

"You need to push the baby out *now!*"

The midwife team positions themselves around me. One set of hands on my left shoulder. One set of hands on my back. Jake is there, too, assisting with strength as a fireman might.

"I'm standing," I say, wanting gravity to help me

They lift me to a standing position on the side of the bed, and I push. One great push with everything I have.

The baby emerges, and we are tilted back into the bed. The midwives' duffle bags open to reveal a mini-ER. I'm expecting to see hot pads, teas, tinctures, or any other mildly groovy hippie supplements inside. Instead, they pull out IV bags, monitors, and other emergency equipment.

And just like that, they place a tiny blue baby on my chest.

"Talk to your baby," they say urgently.

They rub their hands all over him—his arms, his legs, his torso.

"Come on, baby," I say.

I hold him closely.

"Come on, baby. I'm here. I love you. Come on, baby. I'm your mama."

Like magic, he turns from blue to pink.

1998—SAN JOSE, CALIFORNIA

Winning a war sometimes requires dismantling. With time, I've learned I don't mind the process of undoing. I can break the boats, pack the house, leave what doesn't serve us behind. I can take things in pieces with me when I go or take only what I can carry. I can build a new home with what I have. It will always be fine.

The day Todd is born, we are served eviction papers from the rental ranch house in Ferndale. While I was relaxing into life with Jake and our children, thinking the worst of the custody battle was behind us, Cliff had one more trick up his sleeve.

Shortly before Todd's entrance into the world, I traveled to San Jose with Emma for a mandatory court check-in. That same day, Cliff traveled north to Ferndale. We may even have passed each other on Highway 101. Without my knowledge, he had arranged a meeting with our landlords, a couple named Groksen.

If they evicted us, he promised he'd pay them four thousand dollars in cash.

They agreed.

While I sat in traffic on my way to the courthouse, nine months pregnant and annoyed at the legal hoops I was still jumping through after all that time, Cliff and the Groksens entered our home without our knowledge and took pictures of every room. I saw the photos later in a court filing. Pictured were the boxes we were still unpacking, the painting in the hall, Todd's bassinette in the corner of our bedroom, clothes hanging in the girls' closets. The purpose of the act could only have been invasion. A reminder that Cliff would always find some way to exert control over my life.

Cliff probably thinks evicting us without notice will leave me incapable of finding another rental home in Ferndale, which will break our agreement and result in Emma being immediately turned over to him. But we fight the eviction and win. Then the owners claim they want to live in the house now. Though I know this is a lie, there is nothing we can do. Contrary to Cliff's belief, though, after a month in a one-room laundromat motel, we find a new Ferndale house. Still, I have to go back to court to address more contempt charges. I face ten counts of contempt of court, all frivolous and untrue infractions Cliff and his lawyer cooked up to exhaust me financially. I was charged with various minor procedural infractions. Family law courts at that time were overloaded, and contempt motions could be used strategically, especially post-settlement, to regain control or harass. I felt this was done to intimidate me and cause me additional stress. With a newborn baby, our family again shuttles back and forth to court in San Jose, barely able to afford the gas it takes to get us there.

In the courtroom, I leak milk through my blouse. After the long trip, the proceedings last a whopping fifteen minutes. I wait, uncomfortable on a metal chair, while the judge and the lawyers throw legal jargon back and forth at each other across the overly air-conditioned room.

Now that my company is kaput, my plan is to file for bankruptcy. I have no income, but I am still on the line for Cliff's attorney fees, which I have no way of paying. Bankruptcy will get me off the hook for that cost.

Cliff, of course, knows this. He and his lawyer conspire to claim those unpaid legal fees as unpaid child support. Since I've never been ordered to pay child support, I am sure the judge will see right through their ruse.

Through charge after charge, the judge swiftly declares me innocent.

Then, something shocking happens. The judge hands down a guilty verdict on the tenth and final count: failure to pay child support.

I am aghast. I am so obviously innocent on every other count, but this? Perhaps the most flagrantly untrue of them all? Of this I am guilty?

I am sentenced to fourteen days in jail, but the judge suspends the sentence in favor of two years' probation. One misstep in that time, he warns, and I'll go straight to jail.

There isn't much time to process these events. I have to feed my hungry baby.

All the time and energy I've devoted to legal pursuits, I now need to spend rebuilding. There is life to be lived outside of the courthouse, and I want to live it.

CHAPTER TWENTY-TWO

TECH-TONIC SHIFT

"Nothing is softer or more flexible than water, yet nothing can resist it."

Lau Tzu

1999—FERNDALE, CALIFORNIA

Finally, everything seems to settle. The custody fight at long last quieted, Cliff announces his engagement and prepares for a second marriage, and our new house on Main Street in Ferndale starts to feel like home. Emma thrives in Ferndale and starts playing soccer and softball. She still sees Cliff, but not as often, and I don't hear any more of the extremely uncomfortable comments. Now that there is another woman living with him, I trust that this means Emma is safe there.

Not wanting to *ever* go back to court, I decide to focus on being a mom and a housewife and give up on my business life for good. But I still have an incredible amount of shame for letting down my investors, and I can't quite bring myself to do anything with it other than disappear and lick my wounds. Lucky for me, this timing somewhat coincides with the dot-com crash, so I go along with that as the reason for the company's hibernation. Plus, it doesn't seem like anyone blames me personally.

But I hear from investors and attorneys that there is some interest in restarting Wordcasters. Victor Anderson is still around, and other important players, including my marketing guy Mark Marinovich, are interested in working—for stock rather than salary. When I receive this message, I am starting to feel normal again after Todd's birth. Jake's job at Eel River Sawmill turns out to be on the seasonal side, and he's asked to come in less and less often. It's a good time to dip a toe back in the big water.

My attorney draws up the contracts, and my original investors write more checks. They all believe in me and want me to return to technology. First, though, I am accepted into a weekly fast-track learning series backed by Silicon Valley Bank and the Forum for Women Entrepreneurs. It's a way for me to get reconnected and rebuild my network in order to restart my business. I'm one of only eight women CEOs accepted to participate in a series of weekly mentoring meetings with the CEOs of the top technology support firms in San Francisco and Silicon Valley.

To join, I have to get myself down to San Francisco once a week and be fresh upon arrival, which is not easy. Not only do I have to frequently excuse myself to change out my breast pads, but I also struggle to find clothes from my old life that still fit. I have to somehow get my postpartum body into a nice shirt and slacks or a business dress or suit, all freshly ironed, and have my hair slicked back in a professional style. Just finding something to wear is more work than I expect.

Since our big blue Suburban is falling apart and our Jeep has expired Idaho plates, I can't drive either of them. Thankfully, Ramona gave me her old burgundy beater when she moved to New York. It runs just slightly better than the Suburban but takes less gas. I drive that little beat-up car the five hours down to San Francisco and back up to Humboldt County each week. It shakes radically and barely reaches the speed limit. Plus, it still needs to be registered in our name, but I don't have the money to do it and frankly procrastinate the shit out of the task. I make the drive back and forth in a single day because I have nowhere to sleep in the city and because I'm still breastfeeding Todd. Sometimes I pull over and snooze for a while on the side of the road, especially when the 1:00 a.m. hour hits and it's too dark or wet to see the road through the darkness of the majestic redwoods.

Despite all this, when any of the other workshop participants or leaders ask, I pretend I still live in San Francisco. It's easier than admitting to the real circumstances of my life.

Now that I'm getting back into Wordcasters, Jake is home with the kids a lot more. He shoulders a lot of the parenting work while I drive back and forth and generally regain my strength and my brain power. I'm still recovering, both from the trauma of the custody battle and from childbirth. As I get into the swing of things, I am able to rebuild my confidence, little by little. It's much harder this time around because we live so far away, but a dial-up internet connection allows me to create a little office in the closet of our bedroom. It's right next to the kitchen, and I accomplish quite a lot while still cooking and cleaning and being a mom.

Emma is a student at Ferndale Elementary and has awesome friends, especially her cousins Brittany and Amanda. The three girls form a bold, awesome threesome and spend lots of time at our house, running around town like adorable, spunky triplets; grooving out to the Spice Girls on our lawn; and selling lemonade and cookies to the cars that pass by our distinctive historic Victorian on Main Street.

Todd is too young for day care, but Olivia is a rip-roaring preschooler at Ferndale Children's Center, where, like her sister before her, she is often the center of attention, with her adorable swagger, precocious mind, and whip-smart commentary. While Emma was more contemplative at that age, Olivia brandishes boundless, fearless physical and mental energy. No one is faster on the trike. As she rides, she holds philosophical conversations with her teachers. "Multitasker" does not even begin to describe her. No one can climb a tree higher or fall farther without a scratch, all while yelling "I'm okay, Mom!" mid-air as she drops so I won't worry when her little body inevitably thuds to the ground. If she gets hurt, she knows I might not let her climb again, and her preemptive reasoning astonishes and impresses me.

More and more often, I arrive home from work trips to San Francisco to find Jake listening to Rush Limbaugh on the radio. While sweeping the floor, washing the dishes, taking out the trash, all with a growing Todd asleep in the carrier on his back, Jake keeps one ear tuned in to Rush Limbaugh, whose blabbering reverberates around the house.

It's becoming obvious to me that something about this is not healthy for the kids, and again I feel the tug of guilt for leaving home so often. Jake is constantly enraged at some of the things Limbaugh or one of his callers says. Nearly every day, Jake tries to dial in so that he can voice his agreement with Limbaugh and outrage at the rest of the world. *This is a strange cult*, I think, *and Jake is getting indoctrinated.* We're obviously growing apart, and I know I somehow need to extract the kids from the daily diatribes of this fear-driven, polarizing propagandist spewing hate from our portable stereo system.

"Why don't you let me take Todd for a while?" I say more and more often, trying to slip now one-year-old Todd from his carrier on Jake's back. "It'll be good for him to be held for a while."

Todd, like Emma was, is a quiet and reflective one-and-a-half-year-old. He has already figured out that silence is currency. He knows when to spend it and when to save it.

Jake resists. He seems increasingly distrustful of me. I'm reminded of when Olivia was born. Maybe he's once again feeling cast aside or unimportant.

Some days, Jake puts Todd in the baby backpack for a hike. He enjoys being outdoors with the kids from time to time, and it's not unusual for two or more of them to pile into the Jeep and take off for the afternoon. There's a swampy waterhole in Richardson Park up a hiking trail Jake frequents with both Olivia and Todd. It features a rope swing no one bothered mentioning to me. I find out by having some photos developed. When I flip through the finished photos and spot my four-year-old child, a daredevil's scream frozen on her face, clinging like mad to that rope as she cascades over a bed of thick, green algae, I'm anything but pleased.

"What the fuck?" I rant to him. I don't play with the children's safety. "You let the kids swing on this rickety piece of twine over a pond full of who knows what? They could fall in! They could sink down into that muck and never be seen again!"

When I bring it up, Jake, who is indeed a world-class swimmer and diver, tells me I'm overreacting. He says I'm being overprotective and that I should trust him. We have a fight or two about that fucking swamp.

Eventually, Jake trades in his hikes to the waterhole for solo walks in the cemetery. Ferndale has an incredibly beautiful and historic graveyard on a hill at the west end of town. It's nestled behind a charming craftsman-style church that was converted into a home by a reclusive artist, a man who plays the saw as a musical instrument.

Jake goes to the graveyard to talk to the dead.

"I read their names on the tombstones," he tells me. "And I talk to them and cry. I talk to Tom and Jim, Norma . . . and watch the ocean. It's beautiful up there. I ask them if their families are still alive now and if they have forgotten about them. Do they visit? I ask them if it's all worth it."

"If what is all worth it?" I know the answer, but I want to keep him talking. I want to make sure he won't shut me out and slip into some darker place, alone.

"Life," says Jake.

Days pass. Jake again takes Todd out—for a little fresh air, he says. I feel a strange sense of unease, but I'm glad for the quiet time. There are phone calls to make and emails to catch up on.

But when 3:00 rolls around, they still aren't back. It's time to pick Olivia up from preschool, so close the email I'm drafting and head off to fetch her.

When I arrive, the heads of the school, Rebecca and Renee, both look at me funny.

"What's up?" they ask.

"Just here to pick up Olivia," I answer.

Something is off. They look at each other, then back at me.

"Jake already picked her up. A couple of hours ago . . ."

There's concern on their faces, mirroring the unease I felt earlier. We all know this is odd. Jake never picks Olivia up early—at least, not without telling me about it. Clearly something is wrong here.

There is a knowing women get in moments like this. Your body knows it and your mind knows it, even as your logic center takes over and tells you your body and mind are wrong. Bad things happen to other people, not to me, logic says. It's your consciousness protecting you, helping you not to think the worst. You don't have the tools or the certainty to deal with the worst in this moment, so you allow your logic to overrule your instinct.

I drive the three short blocks home, somber. Emma is home from school, and she starts on her homework while I wait. I can't work. I wait. I can't watch TV. I wait. I can't listen to music. Nervously, I wash some dishes. I make Emma a snack. I calmly pace. I look out the window, and I wait. I can't think of anything to do. Where are they?

As the sun begins to set, I bargain with myself. If they are not home by dark, I'll call the police. What will I say? I can't go looking for them. I don't really know exactly where Jake hikes. If it turns out they're just playing at a friend's house, I don't want to give Jake another reason to call me overprotective. But I know they're *not* playing at a friend's house. Oh, what do I do? What do I do? How much longer should I wait?

I finally make dinner, put on a CD of Paul Simon or something, turn on a few lights. I wait. The sun sinks lower and lower in the sky.

I think of my father, disappearing with baby Bryce into the foggy English night, and I wait.

At 7:00, the front door opens. Olivia strolls into the kitchen with Jake at her heels, baby Todd and the diaper bag in his arms. He hands Todd to me without a word, without even looking at me, and goes into the bedroom, closing the door behind him.

I feed the children, read them a story, cuddle with them on the couch. We watch TV until it's time for bed, then I tuck them in. I've been sleeping in the guest room for a while now, so I find my way to bed and fall asleep. I don't see Jake again that evening.

At 3:00 a.m., I awake with a start. Jake is standing over my bed. It's clear he's in distress, nearly hysterical. I turn on the bedside lamp and sit up. He backs away, crying.

"I have to tell you something!" he screams.

His body crumbles into a corner of the room. He covers his face with his hands and frantically sobs.

"Okay," I say. "Tell me. Tell me what's wrong."

I keep my voice slow and even, trying to calm him.

"It's okay," I say again. "You can tell me anything."

He weighs that information and wrestles with it. Finally, he begins to speak.

"I took the kids to the bridge," he says. "I took the kids, and I was going to drive off. I took the kids to the beach—to the cliff. I was looking for the right spot, but I couldn't find it. I wasn't sure it was high enough. So I drove under the bridge, and I drove into the river. I was trying to kill the kids."

I'm hardly breathing.

"*But*," he continues. His voice changes, like he's about to deliver some good news. "I didn't take Emma. I only took Olivia and Todd. And I took the Jeep, not the good car."

I am aware of my body in time. Every skin cell, every pore gasps at the air, hungry to move forward again, to be in any other moment but this. Jake's hand reaches for me. I have to come up with an idea, a solution, anything. He's begging me to help him solve this, to do something.

I blink away the words flashing neon across my mind. *Murder suicide. Murder suicide.*

Don't make this any worse than it is, I think. *Just keep him calm.*

He's not dangerous now, but he was dangerous earlier today—and he can become dangerous again. He can kill me right now if he wants to. He can kill the kids. He can kill himself, right here, in the house, tonight.

I flit between half-second feelings. I can't seem to complete a thought. Instead, my brain scans all possibilities at once, strategizing. Without spooking Jake, I need to get the kids out of the house. I need to get help.

"Everything's okay," I croon. My voice is calm, sweet. I need him to trust me.

He's still in the corner, crying, shaking his head.

"I think—" I begin, measuring my words. "I think we need some help. Help from someone who knows what to do. Do you think you might want to talk to someone? Talk to someone with me?"

He says yes, then no. He cries some more. He says yes and then no again.

"I don't know who to call, but hey," I say, "I think I saw a phone number in the phonebook. You know, at the front of the book? Where the emergency numbers are? I think I saw a number for suicide prevention. Or mental health, you know? In the phonebook. Do you want to try to call one of those numbers and see if they have any advice for us?"

He says no at first but eventually agrees to it. I dial the number, still sitting up in the guest room bed with Jake slumped in the corner. A sleepy man answers, and I calmly explain that I'm with my partner, who's having a little trouble.

"He's having some really big problems today, and he's very emotional. He wants some help," I say. I ask what we should do next. Can this stranger on the other end of the line help us figure out what to do?

"Can you tell me more about these problems? Is the subject with you now?" the sleepy man asks.

"Yes, he is with me now." I cradle the phone and look at Jake. "Do you want to talk? This guy is a therapist on call." I offer him the phone.

"Okay," Jake says.

He takes the cordless phone and presses it to his ear. I can only hear Jake's side of the conversation going forward, but he seems to answer every question posed to him.

"Well," he says after a while, "I drove my kids to a bridge. I wanted to drive the three of us off so that we didn't have to live any longer . . ."

Then Jake's demeanor changes—quickly. He's visibly agitated, and he starts yelling into the phone.

"You are supposed to help me! You tricked me!"

He hangs up.

He's tucked tightly back in the corner again and wailing as if in pain.

"That was *bullshit!*" he says. "That guy was bored until I told him what happened today. Then he got *excited!* I could feel him get *excited!*"

Jake screams and cries at me, wanting me to understand. His wild eyes tell me he could turn on me any second.

"I should never have told him! He said he is required to report what I said to him! That is not helping me! Did you know this was going to happen? You knew!"

He stands up, still wailing, and leaves the bedroom. I'm frozen on the bed. Jake's loud tears blend into the sirens approaching outside, a surreal melody. It's 4:00 a.m. Through the window, I can see the police cars' flashing lights. All three kids are still asleep upstairs.

When the police knock on the door, I let them in. From here, it's a familiar procedure. They separate us, questioning Jake in the kitchen and me in the living room. I want to be near the stairway so I can hear if the kids get up and also prevent anyone from going up to them, especially Jake.

The officer in charge is a blond, balding guy of medium height. He has a bushy blond mustache.

"Did you provoke your husband?" he asks. "Jake tells us you were bossing him around. Maybe you should try being more domestic. That is, if you don't want this to happen again."

He thinks this is a domestic dispute. He tells me someone will follow up with us later that day.

The officer's voice becomes fuzzy as my mind drifts to the bridge. I see Jake there, parked, the kids buckled in their car seats in the back of the car. Jake is crying, trying to muster up the courage to end their lives. Is he thinking of me? Does he know that I'll harness every drop of positivity to take care of them? Of *him*? Maybe he's remembering that night when I sat at his feet and forgave him for ripping newborn Olivia from my arms. I forgave him so easily—just took his hand and told him it was all going to be okay. Did he trust me, again, to do right by him? Is that what brought him to my bedside in the middle of the night? Is that why he confessed? So that I might, again, forgive him? So that I might wash him clean of what he did?

As the sun rises on a new day, light careening across the floor, Jake makes a pancake breakfast. It's like nothing bad has even happened. We've all just woken up from a very bad dream.

Then the telephone rings. There's a therapist on the line. He's following up, and he wants to go over next steps with me.

"We've got some big things to discuss. Child Protective Services has been alerted."

It wasn't a bad dream after all.

"Is the children's father still in the house?" the therapist asks.

"He is," I answer.

"Okay, I need you to listen closely. He needs to leave the house. Immediately. If he fails to do so, we'll have no choice but to remove the children."

If I allow Jake to stay in the house, I'll be considered as much a danger to the children as he is.

"But I'm not afraid of him," I protest. "He is asking for help. He wants to try therapy."

"Ma'am, you don't understand. You have to remove him immediately. Otherwise, we'll be forced to take the children into foster care."

"I understand," I say. There's no choice but to act.

Now, I have to figure out how to make Jake understand. I tell him what I've just heard.

"Maybe we should let the kids stay with Pam," our children's beloved "surrogate grandma" as she likes to call herself. "You and I can drive down to Redway to that attorney who helped me with the custody thing. He was nice. He is a family law attorney, so maybe he can tell us what to do."

Jake agrees. We drive down to Redway and meet with the attorney, who listens to our story and is very kind, calm, and compassionate throughout. He recommends that Jake admit himself to Sempervirens, a care center in Eureka. They'll help him deal with his emotional issues, and it will look good to CPS that he's voluntarily getting help.

With no other options, Jake has to agree. I drive him to Sempervirens. When CPS calls again to check in, I let them know what we've decided.

Later, I call to see how Jake is settling in and am informed that his admittance is no longer voluntary. He is now being held on a 5150, meaning he's a danger to himself or others. They'll have to keep him for a mandatory seventy-two hours.

If Jake had any sanity left, he's lost it now. He spends his days and nights calling everyone he knows on the landline open to patients in the communal area—me, his parents, his friends. The phone rings and rings and rings.

"I'm being held against my will," he shouts through the receiver. "It's illegal! I should be able to leave whenever I want. I mean, I came here voluntarily. Don't you understand?"

He tells me that he fought with the doctors and that they injected him with something without his consent. It's clear to me that he's where he needs to be.

For so long, I was focused on the custody battle and its effects. Perhaps, all that time, Jake's depression was steadily growing, taking more and more of his mind and soul and self. Maybe he carried it with him to Ferndale, to one house, then to another. Maybe it walked alongside him in the cemetery, made breakfast with him in the morning, rode shotgun when he picked the kids up from school. Maybe it was just too quiet to notice. Until it wasn't.

CHAPTER TWENTY-THREE

TERMS OF SURRENDER

"Be like a tree and let the dead leaves drop."

Rumi

2000—FERNDALE, CALIFORNIA

Jake stays in the hospital for five days. When it's time for his release, I receive a call from the staff asking where he can go.

"We can't release him back to you," they explain. "He'll have to leave the county and promise to stay away from you and the kids. Otherwise, he'll have to stay here."

Jake agrees to maintain an appropriate distance from the kids and me. He has a friend in the Bay Area he can stay with, Chris, if only he can get to the airport in Arcata.

I convince the hospital that the kids are safe with a neighbor, but that I'm Jake's only ride. I promise not to waver, and thankfully, they know I'm solid. I also think they really like Jake. They know he's not bad, just troubled, and that he doesn't want to be the way he was before. So they agree to let me drive him but make me promise that I'll pick him up at the hospital and take him directly to the airport. They also insist on seeing his plane ticket before they'll release him.

We're kind to one another as we drive along down the highway. Jake is thoughtful, reflective, happy to be free, and looking forward to seeing his friends and leaving me to bear the weight of our family for a while. He doesn't seem sad or like he missed us. It's more like he's relieved, if not outright grateful.

"This isn't the way I thought life would turn out," he says.

"Life hasn't turned out," I say. "We aren't even forty yet. I'm thirty-two years old, you're thirty-seven. We are young. Life hasn't turned out."

Jake laughs.

"They let me keep my orange jumpsuit!" he says.

He pulls out a neon orange canvas-like item from his bag. It says "INMATE" on the back. He loves it.

"I'm going to wear it around the campfire at the abalone diving camp," he says.

He's in good spirits—contemplative but good—when he boards the plane. I head home to the kids to start up fresh. Time for a new plan.

B ack in the Bay Area, when he's well enough to consider what I might owe him, Jake tries his hand at extortion. He calls my lawyer, Jim Topinka, as we're finalizing the last round of funding to restart my company.

"I never signed an NDA," he tells Jim. "I could easily call up Susan's competitors and tell them how the Wordcasters technology was built."

I never thought to have Jake sign an NDA. After all he watched me go through with my family and with Cliff, he knew how hard I'd worked and how much I'd lost. Now that I'm the sole caregiver for his children, I wonder how he can sabotage me. How can he sabotage the children?

"I'll take one million dollars or 10 percent of the company," he tells Jim. Jim calls me and tells me this over the phone. I can't believe it.

"What do you want to do, Susan?" Jim asks.

"Call his bluff," I answer.

Jim does just that, and it works. Jake doesn't have the courage or strength to follow through on his ludicrous threats, but the damage has been done.

"By law," Jim says, "I have to tell the investors about this. They could pull the funding."

"Okay," I sigh. "Yes, you have to tell them."

Not one of my investors backs out. They believe in me, even after everything. I am awed by that and deeply honored.

Of course, I never told anyone related to the company, even Jim Topinka, about the details of my troubles at home. I didn't tell them about the custody fight. I was deeply ashamed that my personal life was such a strain on me. I was determined to remain capable.

There's a specific kind of grief when you realize someone saw you not as a person, but as a resource. I kept asking myself what I missed. Was I too generous? Too focused on building the dream to notice the rot? That betrayal didn't just break my trust—it made me question my instincts. For a while, that was the most dangerous wound of all.

When Jake is away for so long from the kids, I buy a camcorder and record them blowing him kisses and sharing little messages with him. At night, after the kids are in bed, I transfer the tapes onto VHS and send them to him in the mail. As I slip each finished tape into its envelope, I wonder how he can justify what he's doing . . . a temporary money grab? Even if he hates me, I am still their mother and the one constant in their life now. I didn't want to be a single mother, and our partnership was very successful until—until I started spending more time on my business than on us.

My mother always acted like everything was perfect even though abuse was happening in the house every day. She, too, came from abuse but never spoke of details. Instead, I watched her relationship with her father in his later years, and he seemed consumed by remorse— but for what? I'll never find out. I won't keep secrets from *my* children. I would have wanted to know the truth about my mother's experiences so that we might have abolished the pattern of abuse. Maybe my children remember when their father tried to kill them and himself. Maybe they don't. Maybe they could feel it even if they didn't understand it. I do not know.

When they are teenagers, I do tell Todd and Olivia about what their father did that day. They listen without asking questions. Many years later, when I tell the kids I'm writing this book, Todd tells me that when I told him and his sister that their father tried to kill them, he believed that I was lying. It's a result I didn't anticipate. It is incomprehensible to me, the damage to our relationship caused by his belief all these years that I lied. I hope, more than anything, to repair that trauma.

After all, Emma was old enough to remember. Jake didn't take Emma because she wasn't *his* to take, he reasoned, but she was a witness to it all. In my own confusion and trauma right after these events, I don't know how to risk bringing up the topic without guidance. Would talking about it without proper training in childhood trauma make everything worse? Where is the line between bad-mouthing the other parent and addressing harmful truths? I worry that I will make a horrible situation worse, so I simply put it off until I *do* know what to do.

In telling them the truth in their teens, I do what I think is best. There are no good options, no easy way to explain why a sweet, handsome man who loved his children would try to kill them and himself. Maybe he wanted to save them from the sadness and pain he experienced on this cruel planet. Maybe the pressure of parenthood was just too much.

While Jake is away from home, I manage the best I can. I feel sorry for what he's going through, but I understand that the best thing for everyone is for him to have that time away—for as long as it takes to feel safe with him again. For the first few months, he doesn't want to see the children at all. He's busy reestablishing himself in the Bay Area, living with his buddy Dan and working as a medical equipment repairman, which he'll continue for the next fifteen years.

When, in 2005, I move back to San Francisco with the kids, Jake will visit them more often and even occasionally stay in my house when I travel for work, bringing the kids to and from school, making dinner, spending time with them. I tell him I went to court and was awarded full custody, which he never questions. In many ways, we'll develop a friendly, working partnership, as long as Jake never feels any pressure. I make sure not to cause any.

I n the quiet after the collapse, I finally had time to look around—and what I saw made my stomach turn. The systems I had placed my faith in were never designed to hold people like me. They were built to reward legacy, capital, and compliance. I had tried to play by the rules while also re-writing them, and I paid for that vision in every way a woman can. Bursting my own tech bubble didn't just break my company—it revealed the truth. I hadn't failed. The system had functioned exactly as it was built to.

Thinking back on all the hurt I've held in my life and the people who've participated—Sal, Cliff, Jake, my own mother—I don't even know how to be as sad as I should be. It's a tragedy, but it is a tragedy *they* have to live with. I feel the hurt, the disappointment, deeply, but I work hard to let it go and look forward. Make room for people who care for me and lift me up. I've done everything in my power to create a healthy, normal life for my children. To allow them to flourish, to dream, to grow, to play. I cannot change the past, and I cannot control anyone else's actions. I am not nat-urally inclined toward empathy and emotional nurturing. But I can show the kids, by example, how to thrive, so that's what I do.

As time passes, however, I come to understand that my children hold the trauma somewhere, as do I. Yes, we were victims, but instead of ad-dressing the children's mental health, I put our sorrows in the past and carried on. I didn't know that the trauma was still buried within each of us. How could I know?

I thought I was protecting Emma. I told myself I was doing everything I could. But years later, I started to see it—how she flinched when someone moved too fast, how she stopped asking for help. She'd learned early that her emotional needs made people uncomfortable, even, now, her mother.

I was so focused on keeping everything together—work, court, safety, groceries—I didn't always see her. Not really. She was growing up in the shadows of my own fight. And by the time I looked up, some part of her had gone quiet for good, her innocence stolen.

Looking back, I wish I'd done a better job keeping the outside chaos from bleeding into our home. Maybe then the kids wouldn't have absorbed so much of what I was trying to outrun.

When Cliff sued me for custody, I let his manipulations get inside me. I didn't know then that I had the power to shut the door—not just legally, but emotionally. I thought reacting meant protecting. But all that did was keep me in the fight mindset and force me to relinquish our peace.

I still take things personally when I feel attacked. But I'm learning. Slower than I'd like—but learning.

After Todd was born, I developed a mantra: It is not my job to set everyone straight or to hold people accountable for their behavior. When it was suggested that I report Cliff's lawyer for false statements to the court, for instance, I rolled the idea over in my mind but came out understanding that her having to live with herself and her actions was already punishment enough. This idea offers me so much relief. I can let go and allow karma to take care of us all.

The search for meaning was never abstract for me. It was lived, day by day, in every decision to show up when I was exhausted, in every moment I stood my ground when it would've been easier to fold. I didn't have the luxury of chasing self-actualization. I had children to feed, ideas to protect, and a life to rebuild. But even in the darkest moments, I believed—on some level—that my story mattered ... that survival was not the whole point. I wanted to transform what I'd endured into something useful, even beautiful. This memoir is that offering. It is my truth. And telling it is my way of making peace with all the versions of me who kept going when there was no other choice.

Letting pain define me would have been the wrong choice for me and for my kids. What we needed most, I thought, was to leave all the bad times behind us. One strength that has served me time and again is my ability to reimagine the future and to pursue that new image with my whole heart. Despite everything, we continued—we continue—moving forward.

CHAPTER TWENTY-FOUR

AVENUE OF THE GIANTS

"I'd rather have roses on my table than diamonds on my neck."

Emma Goldman

2001—FERNDALE, CALIFORNIA

After Jake leaves home, I absorb the comfort of my friends and my community in Ferndale, people who know and love my children and me. A year later, while the kids are with our neighbors Pam and Bud, I'm sulking into my beer, lamenting about a guy I've been dating who prefers booty calls to taking me out in public. I'm at the Palace Saloon with Raquel, one of my oldest friends—my first roommate in San Jose, in fact, who has been like a big sister to me. I'm sitting in a sunny spot of the bar with my head low when a cowboy walks up to me and introduces himself. Raquel sits next to me, watching the conversation unfold.

His swagger and his smile stun me. All the other cowboys in the bar turn to watch him approach. They have a front-row seat to the beginning of our romance.

"Well now," the man says, beer in hand. "Don't you look like a daisy in a field of tumbleweeds? My name's Ben. I'd sure like to know yours."

I raise my head and can't help but feel my lips turn up in a lopsided grin. He's poised for a proper introduction, and I appreciate his gentlemanly approach—and his Western flair.

Ben asks for nothing. Not my number, not a date, not to buy me a drink. In this small town of twelve hundred people, everyone knows where you live. He can find me if he wants to.

When he rejoins his friends, Raquel can no longer withhold her excitement. "Oh my god," she says. "That's Ben Truitt. I would be with him if I weren't with Tim."

Well, that's an awkward way to say if she thinks Ben is a good choice for me.

We're still in an era before texting, so Ben calls me on the phone and asks for a date—a real date to a restaurant. I say yes. He picks me up in his Ford F-250, a truck capable of hauling cattle, and takes me over Fernbridge to Parlato's in Fortuna. Ben's dad owns the swankiest restaurant in Ferndale, so if he wants to take me anywhere as nice, it'll have to be over the bridge.

Inside, Ben greets all the cowboys at the bar in a way that makes me feel I'm on display. Then we head into the dining room, and it's just us. I'm not a bar girl, and Ben knows it.

When we're seated and looking over our leather menus, Ben takes charge and orders a bottle of wine.

"We'll have a bottle of the Sangiovese," he says to the waiter.

This Western cowboy really just said *Sangiovese*. I look up from my menu, and Ben winks. I'm in love in that moment. I've been casually seeing this other booty call guy—Tucker, a classic player who chases all the single moms. Then one night, after we've been dating a couple of weeks, Ben stops by the house and tells me, with the greatest respect, kindness, and love, "I'll take you on. You and the kids. But you have to stop seeing that other guy."

"Okay," I say. And that's that.

Ben raises my kids with so much love. He teaches us all what love is. It's perfect, having a reliable, caring man in our life. Someone who doesn't want anything more than to love us. And that's what he does.

Ben takes us to his ranch, high on a ridge near a town called Blue Lake about an hour away from Ferndale. He and his business partner own two hundred head of cattle. Every spring, they brand calves in the traditional way—with ropes and horses. My kids learn to eat Rocky Mountain oysters, ride horses, build fences, and chop wood for the winter. We hop in my old Idaho Jeep and drive out to the ranch's five barns, then patrol the ridge at sunset. In wintertime, the entire ridge is covered in snow, so we work all summer long to keep the firewood shed full.

Ben introduces me to his dad, Curley, who asks me to help him transition his restaurant from paper to QuickBooks. I have no idea how to do that, but I agree. In the process, I fall in love with the food business and eventually become the general manager of the restaurant and catering business. Curley's Grill thrives under my watch.

I love this life. The ranch suits me, and I still get to fulfill my business cravings by managing the restaurant's forty employees. In this old-fashioned community, I even reconsider my aversion to marriage. Ben wants to have a baby, and I'm open to the idea.

Then, as naturally as anything else out here in the countryside, Ben proposes to me, and I accept. Once we're engaged, we start looking at more ranch property to lease or to buy. We want to make a "homeplace," as Ben calls it. Meanwhile, he continues to manage the multiple ranches he leases, running cattle with other ranchers across the inner and outer wilds of Humboldt County. I settle into helping out at the ranch and focusing on the children's growing needs and sports as they mature into their preteen years.

My mother and Sal love the idea of a real, authentic cowboy in the family. They fawn over Ben and invite us to join them in their newly built country home in downtown Ketchum, Idaho. Ben and I take the kids to ski there in the winter and play golf in the summer. I even start to enjoy hanging out with my mother and Sal and build a positive relationship with them.

I make the deliberate choice to sincerely dedicate the next ten years to working toward forgiveness with my mother and Sal. Perhaps, with effort, I can lead them into becoming the family I wanted so badly as a child—one in which I'm respected for who I am and loved with eyes wide open. With Ben in the family, the plan seems to be working. My mother and Sal dote on me and the kids, happy to share in this life we're building in Ferndale.

One evening, Ben and Sal share nightcaps and cigars on the garden terrace overlooking my parents' pool.

"I am sure Susan has told you about her childhood and what an asshole I am," Sal opens, timid.

"No, Sal," Ben replies. "She really hasn't said anything like that to me. She has mentioned some trouble growing up, but she has no ill will or hard feelings. When she speaks to me about you, she calls you her dad."

Sal, Ben recounts to me later, goes on to tell him how lucky he is. "Susan is strong, honest, true, and beautiful. Keep her safe," he says.

Of course, Ben *does* know some of the highlights from my childhood, but I never want to define myself by the past. Still, it's moving to me to hear, even secondhand, Sal acknowledge his behavior and express regret over it. It's a huge step for the Mr. Caputo I used to know, and it must come straight from his heart. Turns out he does have one after all. Of course, I would prefer to hear the words in person, but I decide to accept this as enough. It's all I need to be able to heal. I no longer feel the need for validation, apologies, recovery, or change. I decide to leave the past in the past and build on the future—with all its promise—for me and for my family.

Ben and I are happy for the next few years. Eventually, though, Ben's whiskey-drinking and cowboying lifestyle become too much for me. When he drives the children, drunk, in his F-250 and winds up in a ditch, I know something has to change.

At first, I consider becoming an alcoholic myself. After all, I've seen so many happy couples posted up at the bar night after night. But I can't do it for long. I grow tired of drinking at the restaurant until closing every night, then stumbling across the street to the Ivanhoe or the Palace Saloon to slur words with other drunk parents and cowboys and dairymen, all our friends in this small town we love. Toward the end, I stay home and let Ben run nights. But when he comes home more and more belligerent and so intoxicated that he falls onto the sofa or the floor in the dark, breaking our paned-glass window on his way down, I realize I don't even recognize him when he is like that. I love Ben, and he loves us, but his behavior isn't okay for children.

Emma is about to enter high school—she craves art, books, and cafes, and she wants to feel the pulse of the city again. Livvy is lightyears ahead of the other kids in this small town, outsmarting all her teachers and babysitters, bringing them to their breaking points. Olivia doesn't enter a room; she collides with it. She needs focus and a few guardrails. What independence! What power she embodies! I have the profound sense that she is going to need more stimulation than this town has to offer.

Todd is entering first grade, and I know that Ferndale is precisely where he needs to be—on the ranch, feeding the dogs, riding horses, repairing fences, cutting wood for the winter with men—but the girls are both at a pivotal age. I vow to return to Ferndale with the kids every month and maintain the friendships and community so valuable to us. And we do.

Ben and I remain the dearest of friends for a long time afterward. He goes on to meet his future wife, Margot, and they welcome a baby boy soon after. Even then, the love between us never disappears. We simply understand it has no place in our new lives. Out of respect for his marriage and growing family, I quietly step back from our friendship. To this day, my children maintain a loving bond with Ben and consider him the most important father figure in their lives—the one who showed them their value with his unconditional love. I will be forever grateful to him for the gifts of love he showered on us when we needed it the most.

In the end, the love story that defines my life does not involve any man. It's the love I feel for my children. And isn't that perfect? After seven years in Ferndale, the kids and I go back to San Francisco. I get a job in Napa Valley, and we settle back into our San Francisco life.

Emma, Olivia, Todd, and me—we start over again.

The world I inherited was made of broken systems and frayed promises. But somehow, in the gaps between collapse and creation, I found space to build the kind of safety I once dreamed of.

I never got the patent. I never got the billion-dollar exit. I didn't get the courtroom justice I begged for.

But my children? They got a mother who never stopped choosing them.

They got a life shaped by a woman who walked through fire, barefoot and uninvited, and still showed up with bread in hand.

Let the critics say what they want. That I was emotional. That I was difficult. That I was too much, or not enough.

I know the truth. I was shaped by a system that never saw me clearly. This is a story about using the broken pieces to build something whole.

They gave me ruin.

I returned with form.

EPILOGUE

"If your everyday life seems poor, do not blame it; blame yourself, for you are not enough of a poet to call forth its riches."

Rainer Maria Rilke

1985—SAN JOSE, CALIFORNIA

Three days after I graduate from high school, I have to get a cat, fast. My life in Hawaii with my father didn't turn out the way I'd hoped. My father and I barely knew each other, and we struggled financially. There were days I begged neighbors for an egg just to have something, *anything* to eat. Finally, I decided I could survive the last few months of high school with my mother and Sal, and I flew back to California.

When graduation came, Sal let me know that I was no longer welcome under his roof. I was only seventeen years old, but now that I was finished with high school, he considered his obligation to me fulfilled.

I move into a room in a house in Almaden Valley, not far from the car wash where I work. My landlady, Sudha Choudary, a traditional Indian woman, rents two rooms in her house, where she lives with her four-year-old daughter, Kajal, and another young woman, Raquel, and her young daughter, who is also named Raquel. When Sudha interviews me for the room for rent, she asks if I have any pets.

In that moment, my seventeen-year-old impulsivity answers for me. "Yes, I have a cat."

I haven't thought this through, but I definitely want a cat.

"What is the name?" Sudha asks.

Ummm.

She must see me squirming. I make up a name, then try to change the subject.

When I pass the interview, I immediately start looking for a cat. I find a kitten at a pet store and hope Sudha won't notice that, unlike my pretend cat, it's only a couple of months old.

The two Raquels and I quickly become close. They'll still be like family to me forty years later. I learn everything about being an adult from big Raquel, who's five years older than I am. Raquel loves Gianfranco Ferré perfume, so I love Gianfranco Ferré perfume. She loves tank maxi dresses, so I love tank maxi dresses. The two Raquels have a huge bedroom next to my small one, with room for a couple of sofas and a TV. Raquel tapes *All My Children* every day on her VCR, and we watch it together in her room in the evenings after work. I feel safe, loved, and very much at home there.

At work, I start as a lowly vacuumer, a dirty, dusty, and very noisy job that requires me to spend hours with my head bent down in the footrest of other people's cars. Car after car, hour after hour, I vacuum dirt, barf, and old french fries and throw out people's garbage. I end every day with a sore back and a face covered with a fine layer of shoe dirt. I blow my nose, and mud comes out.

Soon enough, though, I'm accepted into the group of kids who work there, and I'm promoted to another area of the line: windows! Windows comes with its own issues, mostly being wet all the time. I'm not crazy about the wet socks and pruned feet, but at least it's a job I can do standing up.

The car wash crew works together and plays together. It's a family. Most of the kids are coupled up. Many of our managers have made this job a career. Rising through the ranks from vacuumer to general manager is commonplace in the Classic Car Wash world.

The owner, Frank Dorsa Jr., is the heir to Eggo Waffles. His father invented the breakfast staple, and Frank Jr. built a chain of car washes with the fortune. I find this intriguing and incongruous. So that's how you get rich and create more opportunities for yourself—invent something like Eggo Waffles.

I quickly move up to cashier, which is great fun, especially on busy weekends when the line through the gift shop can run twenty or thirty customers deep. When it's that busy, two of us work behind the cash register, taking turns ringing up and running credit cards, taking cash, printing receipts, redeeming coupons, and coordinating with the gift shop ladies who sell fur coats, fine jewelry, and specialty items of all kinds. Time flies fastest on busy weekends, which I appreciate.

We wear cute uniforms—ruffled, putty-colored aprons with large pockets and a wide band to tie into a fat bow in the back over a puffy-sleeved white blouse. We can dress it up with our own pants. I usually wear loose pedal pushers with a pink and gray paisley print with ruffled cuff socks turned over at the ankle and White Mountain lace-up oxfords. I appreciate wearing a cute, functional, comfortable, and high-quality uniform to work each day.

After work, the crew meets up and goes out to dinner at Pedro's, an upscale Mexican restaurant in Los Gatos. Sometimes we catch a movie or go to a dance party at one of the older employees' houses or get someone to buy us beer or Bartles & Jaymes wine coolers at the AM/PM mini-mart across the street. Then we all caravan deep into the hills of old Almaden to the spillway, where we slip through a broken fence and climb in the dark, beers in hand, and walk along the narrow ridge of the waterway to a huge irrigation culvert. There, we feel like we're way out in the country. We drink and play games like Bizz Buzz as the dark water pours out of the culvert under our feet. We wonder aloud how many homeless campers live here, just out of sight, how many mountain lions. There's always the danger of falling down between the concrete walls and into the crushing water below. But we never do. We're young and invincible.

I hope it will go on like this forever, but car wash life begins to lose its luster by the time I hit nineteen. By now, most of the crew has moved on to better things, like college or retail jobs in malls. The boys I used to think were cool pick up cocaine habits. I try it twice but absolutely hate the intense paranoia and fear it gives me, like being in a lucid nightmare. I'm not a pot smoker either, and while I do enjoy my wine coolers, the daytime talk show routine of druggies wasting days upon days in a dirty, dark room eating Kraft mac and cheese doctored up with sliced hot dogs is not for me.

It's time to move on.

As I read through the classified ads, I wonder about the type of adult I'll become. I yearn to understand myself—my work ethic, my abilities, my capacity to learn complex things. I know, innately, that I can learn, design, modify, and create things from scratch if I need to. I might not instantly rise to any leadership position—the world sees me as a skinny blonde nobody, after all—but I can create my own circumstances.

So much of my life is still in shattered pieces from my childhood, but I'm finding my own way in my work life, establishing my abilities and learning the boundaries of my skills. My personal life, on the other hand, is nonexistent. I've never had a real boyfriend. I don't see any of my high school friends anymore. I was too embarrassed about being the abused kid in the neighborhood, so I just left them all behind.

Being unloved made me feel unlovable. I realize I'm still looking at the world from the perspective of a child, beaten and berated. How can I create an adult Susan powerful enough to change that?

As I feed my cat and watch her happily devour the meal, I dream the two of us into the future.

I don't belong to a family, I think, *but that's okay. I can grow up and be alone forever and be just fine.*

The cat rubs sweetly at my legs. I reach down to scratch behind her ears.

Or, I think, *I'll make my own family someday.*

END OF BOOK ONE

AUTHOR'S NOTE

T his is a personal story, told through my own eyes—but it lives alongside the stories of those I love. While I wrote it from my own perspective, I recognize that some who appear within these chapters carry their own memories and truths, which may not align with mine, and I hold that complexity with respect.

To protect privacy, names and identifying details have been changed. In some cases, characters are composites drawn from more than one person or experience for clarity and narrative flow. Dialogue has been reconstructed from memory to reflect the emotional truth of a moment, not as a verbatim transcript. While based on real events as I experienced them, this is a memoir, not a comprehensive or documentary record. Some details are shaped by the understanding I had at the time.

Certain businesses, institutions, and cultural references in this book are included as I understood them then. Notably—Lockheed Aircraft Corporation, a major U.S. defense contractor—my view at the time did not reflect the problematic political, social, and ethical realities I recognize today. Their inclusion here is not an endorsement, but part of the historical and personal landscape in which these memories took place.

I wrote this book for mothers and children who, like I once did, endure betrayal and confusion in silence. Acts of moral corruption too often go unchecked in families, courts, and culture. Too many women have been discredited for trying to protect their children. These women are not difficult. They are brave. And they deserve to be heard.

It took time and distance for me to write this book, and I did so with full awareness of the pain it carries. Some in this story caused serious harm. I hold compassion for each of them, though it may not be visible on the page. This book is not for them—it is an attempt to end destructive cycles and offer the witness I needed when I had no words for what was happening.

That is what I've tried to do. I am still trying.

ACKNOWLEDGMENTS

Thank you to everyone who held me together while I unraveled this story. Specifically, thank you to fellow memoirists Ellen Goldstein and Melanie Thomas Armstrong—steadfast supporters, fierce creatives, and chosen sisters. Together, we built what we needed: writing retreats on the remote beaches of the Caribbean and the Pacific coast of Oaxaca. I treasure your friendship—and can't wait for our next writing workshop disguised as a vacation.

To my childhood friend Greg Fox—thank you for your sharp editorial instincts, creative ideas, and lifelong friendship. Thank you for validating my experience. While others on our block had idyllic childhoods, you saw what I lived through. When you told me mine was the house every kid avoided, something lifted—I'd long believed I was the reason they stayed away. That kind of honesty gave me permission to trust my memory and the courage to write it down.

Thank you to fellow authors Kate Coburn, Juliane Bergmann, Brenda Smith, and all the members of the ad hoc memoir groups we created or joined over the past five years. Your feedback, presence, friendship, and care for me helped shape this book.

To the team at Ballast Books—thank you for your stewardship and your belief in this story. Journey and Lauren, it's been a joy to work with you. Deep gratitude to Kat Dixon for helping me see the form in the mess of words I sent you and for guiding the shape of the manuscript. And most of all, thank you to Emma Sherk—for understanding my voice, honoring my story, and helping me tell it the way it needed to be told.

A quote pasted to my wall kept me going and helped me stay true to what was possible when I wrote. I scribbled down Cheryl Strayed's wisdom she delivered at a Kripalu Center weekend workshop—these words helped me remember what I set out to do and to not give up until I got it right:

"Thank you, dark teacher, look at what I did with what you gave me
—I made something beautiful."

www.ingramcontent.com/pod-product-compliance
Lightning Source LLC
Chambersburg PA
CBHW031455120626
46545CB00005B/1620